The Howard W. Sams
Troubleshooting & Repair Guide to
TV

The Howard W. Sams
Troubleshooting & Repair Guide to
TV

By
The Engineering Staff of
Howard W. Sams & Company

***PROMPT*®**
Publications

An Imprint of
Howard W. Sams & Company
A Bell Atlantic Company
Indianapolis, IN

©1996 by Howard W. Sams & Company

FIRST EDITION, 1996

PROMPT® Publications is an imprint of Howard W. Sams & Company, a Bell Atlantic Company, 2647 Waterfront Parkway, E. Dr., Suite 300, Indianapolis, IN 46214-2041.

All rights reserved. No part of this book shall be reproduced, stored in a retrieval system, or transmitted by any means, electronic, mechanical, photocopying, recording, or otherwise, without written permission from the publisher. No patent liability is assumed with respect to the use of the information contained herein. While every precaution has been taken in the preparation of this book, the author, the publisher or seller assumes no responsibility for errors or omissions. Neither is any liability assumed for damages resulting from the use of information contained herein.

International Standard Book Number: 0-7906-1077-9

Library of Congress Catalog Card Number: 96-68278

Acquisitions Editor: Candace M. Drake
Editor: Natalie F. Houck
Contributing Editors: Tim Clensy, Bill Fink, Rida Scott, Dave Sullivan
Assistant Editors: Pat Brady, Karen Mittelstadt, Jim Surface
Illustrations: Tim Clensy, Walt Stricker, Terry Varvel, the Howard W. Sams PHOTOFACT Department
Typesetter: Natalie Houck
Cover Design: Phil Velikan

Additional Illustrations Courtesy of: Jim Hollomon, Jr.

                                          Master Publishing, Inc.
                                          522 Cap Rock Drive
                                          Richardson, TX 75080-2036

                                          Van Valkenburgh, Nooger & Neville, Inc.
                                          33 Gold St.
                                          New York, NY 10038-2815

Trademark Acknowledgments: PHOTOFACT® is a registered trademark of Howard W. Sams & Company. All terms in this book that are known or suspected to be trademarks or services have been appropriately capitalized. PROMPT® Publications, Howard W. Sams & Company, and Bell Atlantic cannot attest to the accuracy of this information. Use of a term in this book should not be regarded as affecting the validity of any trademark or service mark.

Printed in the United States of America

9 8 7 6 5 4 3 2 1

Table of Contents

Introduction 1

Chapter 1
TELEVISION BASICS 5
   How TV Produces a Picture and Sound 5
   Video Basics 7
   Sound Basics 8
   Using a Block Diagram 8
   Receiving the Signal 9
   Increasing the Signal's Strength 12
   Separating Video and Audio 13
   Processing the Video 13
   Processing the Audio 14
   Processing the Sync Signals 14
         Horizontal Sync Signals 15
         Vertical Sync Signals 16
   Producing the Picture 16
   Quiz 17
   Key 18

Chapter 2
WORKING SAFELY 19
   General Guidelines to Follow Before Returning the Repaired Receiver 20
   Testing for Cold Leakage Current in Receivers with Isolated Ground 21
   Testing for Hot Leakage Current 21
   Avoiding Electrical Shocks When Servicing High-Voltage Circuits and CRTs ... 22
   Avoiding X-Ray Radiation and High-Voltage Limits 23
   Avoiding Fire Hazards 24
   Working with a CRT 24
   General PHOTOFACT Safety Symbols 26
   General PHOTOFACT Safety Precautions 26
   Quiz 29
   Key 30

Chapter 3
THE BASICS OF TROUBLESHOOTING 31
   Basic Equipment for Troubleshooting Televisions 32
   General Techniques for Servicing Televisions 35
   Troubleshooting Methods 37
         Signal Injection and Signal Tracing 37
         Analyzing Circuit Voltages 40

       Measuring Resistance 41
       Substituting Parts 41
   Testing Components 42
       Locating Defective Transistors 42
       Locating Defective Diodes 43
       Locating Defective SCRs 43
       Locating Defective ICs 44
   Surface-Mount Technology (SMT) 44
   Servicing Solid-State Devices 46
   Digital Circuits 48
   Helpful Diagrams 48
       Parts List 48
       Placement Chart 49
       Schematic 50
   Quiz 53
   Key 54

Chapter 4
TROUBLESHOOTING POWER SUPPLIES 55
   Standby Power Supplies 56
   Conventional Power Supplies 56
       Transformer 57
       Rectifier 60
           Forward-Biased Diode (VF) 61
           Reverse-Biased Diode (VR) 61
           Half-Wave Rectifier 61
           Full-Wave Rectifier 63
           Bridge Rectifier 63
       Filtering 64
       Capacitor 65
       Troubleshooting Conventional Power Supplies 66
   Locating Power Supply Problems 67
   Regulated Power Supplies 68
       Zener Diode 70
       Series-Pass Feedback Voltage Regulator 72
           Troubleshooting a Series-Pass Feedback Voltage Regulator 72
   Switched-Mode Power Supplies (SMPS) 72
       Switched-Mode Power Supply Operating Principles 73
       Regulators in Switched-Mode Power Supplies 75
       Input Voltage 75
       Control Element Switch 75
       Catch Diode 75
       Inductor 76
       Filters 76
       Troubleshooting Switched-Mode Power Supplies (SMPS) 76
   Scan-Derived Power Supplies 77
       Troubleshooting Scan-Derived Power Supplies 78
   Quiz 81
   Key 82

TABLE OF CONTENTS

Chapter 5
TROUBLESHOOTING VIDEO CIRCUITS 83
    Troubleshooting the Video IF Amplifier 85
    Troubleshooting the Video Detector 87
        Troubleshooting Symptoms 88
    Troubleshooting the Video Amplifier 89
    Audio/Video (A/V) Switching 92
    Comb Filter 92
    Delay Line 92
    Picture Adjustment Controls 92
        Peaking Coil 94
        Sharpness Control 94
        Contrast and Picture Controls 95
        Brightness Control 96
        Vertical and Horizontal Blanking 96
    Last Video Amplifier 96
    Chroma Processing 97
        Automatic Kine Bias (AKB) 98
        No Color 98
        Color Too Intense 98
        Losing One Color 99
        One Color is Incorrect 99
        All Colors Are Incorrect 99
        Loss of Color Sync 99
        NTSC Color System 100
    Picture-in-Picture (PIP) 101
        Troubleshooting the PIP 101
    How a CRT Works 102
        Troubleshooting a CRT 104
    Static and Dynamic Convergence 105
        Convergence Adjustments 106
    Quiz 109
    Key 110

Chapter 6
TROUBLESHOOTING TELEVISION AUDIO 111
    Processing Audio Signals 112
    Sound IF Amplifier 113
    Audio IF Detector 114
    Stereo 114
    Surround Sound 115
    Audio Amplifier 115
    Troubleshooting Symptoms 116
        No Stereo 116
        No SAP 117
        Troubleshooting Example 117
        No Sound 118
        Weak Sound 118

   Squealing or Whistling Sounds ... *119*
   Humming and Putt-Putting Sounds ... *119*
   Crackling or Popping Sound ... *119*
   Distorted of Intermittent Sound ... *119*
  Quiz ... 121
  Key ... 122

Chapter 7
TROUBLESHOOTING DEFLECTION CIRCUITS ... 123
  Sync Separator ... 125
   Troubleshooting the Sync Separator ... *126*
  Automatic Phase Control (APC) ... 129
   Troubleshooting the APC ... *129*
  Horizontal Oscillator ... 131
   Troubleshooting the Horizontal Oscillator ... *131*
  Horizontal Deflection Circuits ... 132
   Horizontal Output Circuit ... *133*
    Horizontal Scanning ... 134
    Triggering a Scan/Retrace Sequence ... 135
   Pincushion Correction Circuit ... *136*
   Troubleshooting a Horizontal Output Circuit ... *137*
  Vertical Deflection Circuits ... 139
   Vertical Oscillator ... *139*
   Feedback in Vertical Circuits ... *141*
   Pincushion Circuit ... *141*
   Vertical Output Circuit ... *143*
   Troubleshooting the Vertical Deflection Circuits ... *144*
  Quiz ... 147
  Key ... 148

Chapter 8
TROUBLESHOOTING HIGH-VOLTAGE CIRCUITS ... 149
  Flyback Transformers ... 151
   Troubleshooting a Flyback Transformer ... *151*
   Shorted Flyback Transformer ... *152*
   Noisy Transformer ... *153*
   Excessive Voltage Level ... *153*
   Replacing a Flyback Transformer ... *153*
  Boost Voltage ... 154
  X-Ray Protection ... 154
  Focus Circuits ... 155
  Automatic Brightness Limiter (ABL) Circuit ... 157
  Troubleshooting High-Voltage Circuits ... 157
  Quiz ... 159
  Key ... 160

TABLE OF CONTENTS   PAGE ix

Chapter 9
TROUBLESHOOTING TUNER CIRCUITS 161
    RF Amplifier 163
    Oscillator 164
    Mixer 165
    Tuner Types 165
    Troubleshooting Tuner Circuits 166
        Drifting 167
        Strong Raster with No Picture or Sound 167
        Strong Raster with Weak Picture and Sound 167
        Weak Raster with Weak Picture and Sound 167
        Weak Raster with No Picture or Sound 167
        No Raster, and No Picture or Sound 168
    Replacing a Modular Tuner 168
    Quiz 169
    Key 170

Chapter 10
TROUBLESHOOTING SYSTEM CONTROL CIRCUITS 171
    Reset Circuit 173
    Audio Mode Circuits 175
    Channel Memory (RAM) 175
    Tuner Control Circuits 177
    Clock Functions 177
    On-Screen Display Circuits 178
    Closed Caption Circuit 179
    Picture Control Circuits 180
    Remote Control and Keyboard Circuits 180
    Power Supplies 182
    General Troubleshooting Techniques 182
    Quiz 185
    Key 186

Appendix A
TROUBLESHOOTING SYMPTOMS 187
    General Television Problems 188
    Picture Symptoms 194
    Audio Symptoms 202

Appendix B
PHOTOFACT 207
    Quiz 221
    Key 222

Bibliography **223**

Index **225**

Introduction

Introduction

This book has come into being out of pure necessity. Over the past few years, the number of television repair centers and knowledgeable repair technicians has dwindled due to manufacturer and consumer attitudes toward televisions as "disposable." For the past fifteen years or so, many people have acquired the opinion that it's just as expensive to get a TV repaired as it is to buy a new one, so why not just toss out the old TV? This isn't necessarily a good idea.

The expense of TV repair has climbed, there's no doubt about that. However, while part of this has to do with the fact that there are repair centers that deliberately overcharge their customers, a lot of it is the result of a lack of individuals (including technicians) who know how to repair a TV.

You can blame it on the lack of repair information available, or blame it on the fact that technical schools and repair centers have not placed an emphasis on learning TV repair because of the TV's "disposability." Whatever the case, televisions do not have to be tossed out the minute they develop glitches or problems, and they do not have to be expensive to repair. In our ever-changing global economy, buying a new television to replace an old one is not always an option. If an individual has the proper information and tools at their disposal, then repairing a TV becomes simple and very economical. With a minimal amount of technical knowledge, you can learn to repair your own TV with the help of this book.

This book was written using the years of experience accumulated by Howard W. Sams & Company technicians while creating PHOTOFACT, which is technical data and schematics on televisions that have been created since the end of World War II. For 50 years, Howard W. Sams & Company has provided the most timely and state-of-the-art TV technical data available, and has introduced many techniques to help service technicians and electronics hobbyists.

Following World War II, electronics manufacturers' time, effort, and resources were directed toward meeting the increasing demand for new equipment and parts, and toward trying to integrate new electronic technology developed during the war. As a result, most manufacturers stopped producing service and replacement parts information. Individuals found it increasingly difficult to find the information needed to repair certain electronic components and technical equipment, including TVs—not unlike the situation that exists today. It soon became apparent that something needed to be done to make service information readily available to the professional electronics

INTRODUCTION

technicians, as well as to individual hobbyists who preferred to make their own repairs.

To produce readily available service and replacement parts data, Howard W. Sams rented 5,000 square feet of space in Indianapolis, Indiana in 1946, hired 12 employees, and began producing PHOTOFACT.

PHOTOFACT is designed to provide as much detailed information as possible for easy servicing of individual television sets. Contained in each PHOTOFACT folder is technical data and schematics covering IC functions, important parts information, miscellaneous adjustments, parts lists, safety precautions, schematics, wave forms, voltages, troubleshooting tips, and more. Throughout the changes that televisions and the industry have experienced over the years, PHOTOFACT has remained the most up-to-date and accurate source of TV service data available. PHOTOFACT schematics are used throughout this book to provide you with the most comprehensive and user-friendly information possible for the television part that you are troubleshooting. Appendix B also provides detailed information about PHOTOFACT and how to use it.

The Howard W. Sams Troubleshooting & Repair Guide to TV contains all the information the novice TV repair technician needs to service and repair all models and makes of televisions. Chapter 1 covers *Television Basics*, including how TVs produce pictures and sound, how they receive signals transmitted from broadcasters, and more. This information is useful because it will be easier for you to service a TV once you understand exactly how it works. Chapter 2, *Working Safely*, covers basic safety tips and safety problems commonly encountered in TV repair. Chapter 3 covers *The Basics of Troubleshooting*, outlining general techniques and troubleshooting methods. Chapters 4 and 5 explain *Troubleshooting Power Supplies* and *Video Circuits*, including CRT and PIP. Chapter 7 covers *Troubleshooting Deflection Circuits*, which control the horizontal and vertical scanning that produces pictures on the TV screen. *Troubleshooting High-Voltage Circuits* is described in Chapter 8, and *Tuner Circuits* are covered in Chapter 9. Finally, *Troubleshooting System Control Circuits* is featured in Chapter 10, which also covers closed caption, remote control and channel memory (RAM) circuits. Each chapter includes a brief quiz at the end as a summary to the information covered, and Appendix A contains a more extensive test for students or for anyone wanting to test their knowledge of TV repair.

Acknowledgments go out to everyone who had a hand in this book by suggesting additions, changes, topics and more, and by implementing them for inclusion in this book. Many thanks to Rida Scott and Avalon Group; to Dave Sullivan; to Tim Clensy, Bill Fink, and the Howard W. Sams PHOTOFACT department; to Evelyn G. Turner and Jim Hollomon, Jr.; to Jim Allen and Gerald Luecke of Master Publishing; to Terry Varvel, Walt Stricker and Phil Velikan of the Howard W. Sams art department; to Jim Surface; to Candy Drake, Natalie Harris and Karen Mittelstadt of PROMPT Publications; to Doug Curran; to Jerry Butcher; to Peg Leeds; to Richard Hauser and Damon Davis; and to all of the electronics distributors out there who have been requesting this book for a long time. Many thanks.

Chapter 1
TELEVISION BASICS

Chapter 1
Television Basics

A television performs three tasks: it shows a picture (video), it produces the sounds (audio) that accompany the picture, and it synchronizes the picture it produces with the picture that is transmitted. A picture without sound is unacceptable, and sound without a picture is like watching a radio. If the picture on the screen is not synchronized with the picture that is transmitted, the resulting picture is chaotic. This chapter describes the basic operation of a television, and how it receives signals, converts signals, and produces both picture and sound.

How TV Produces a Picture and Sound

A picture is a series of tiny squares. Using black and white as an example, the squares range from black to white with many gray tones in between to add definition. When you magnify a black and white photograph, the small squares that define the picture become apparent, as shown in *Figure 1-1*.

The signal a television station transmits also is made up of tiny squares of light and the spaces between the squares. You can see the tiny squares on the television screen when you tune to an unused channel. At the television station, the tiny squares that define a picture are converted to electrical

Figure 1-1. The circle shows a close-up of the tiny squares that make up a TV picture.

CHAPTER 1: TELEVISION BASICS

signals. Accompanying the electrical signals is all of the information about the squares of light, including their positions and their intensity. This information is used by the receiver that converts the signals to a picture to duplicate the transmitted picture. The more tiny squares there are in a picture, the better the picture quality or resolution.

Also, the television receiver must be able to stay in step with the television station's transmitter. The television camera scans a scene much like a person scans a page of printed material. Starting at the top-left corner, a person scans the first line left to right. Then, when the first line is complete, the person's eye moves down one line and back to the left side of the page. This pattern continues until the page is finished.

To ensure that the receiver stays in step with the transmitter and duplicates the transmitter's left-to-right and top-to-bottom scanning pattern, a series of signals or timing signals called synchronizing (sync) signals is sent by the transmitter, along with the picture and sound signals. Using the sync signals, the receiver can stay in step with the camera's scanning pattern. Sync signals can be divided into two sets of instructions. The right-to-left instructions are called horizontal-sync signals. The bottom-to-top instructions are called vertical-sync signals. If there are problems with the sync signals, the picture on the television appears to flicker, tear horizontally, or roll vertically.

Reducing flicker is discussed in the next section. Horizontal tearing and vertical rolling are discussed in the section *Processing the Sync Signals*, later in this chapter.

Video Basics

There are 525 horizontal lines in one television picture (frame), with 285 squares of information on each line. In the U.S., a television station transmits 30 complete frames every second. Transmitting only 30 frames per second causes the picture to appear to flicker. This flickering is reduced by interlacing the 525 horizontal lines. Starting at the top of the screen, the electron gun in the picture tube (cathode ray tube or CRT) evenly scans 262+ lines down the screen, shown as solid lines in *Figure 1-2*. Then, the electron gun goes back to the top-left of the screen and scans another 262+ lines between the first lines. The second set of scan lines are shown as dashed lines in *Figure 1-2*. Interlacing the lines reduces the flickering and produces much the same effect as 60 complete pictures-per-second.

A set of 262+ lines is called a field. It takes a set of two fields to produce one complete frame.

Figure 1-2. Interlaced scanning of the horizontal lines of a TV picture.

If the television is not producing a picture, it produces a scanning pattern. This pattern, called a scanning raster or snow, appears as a series of horizontal white lines even though you cannot see the defined lines. The snow pattern is important to show that radio frequency signals are being received. The radio frequency signals carry the video, audio and sync signals. You can reproduce a scanning raster by disconnecting the television's antenna and tuning to an unused channel. If the snow pattern is absent, there is a problem with one of the early stages of reception, such as the tuner or the IF stages.

In addition to the blanking pulses used to blank retrace lines at the end of the sweep cycle, the newer televisions have a built-in blanking circuit which blanks the raster if there is no signal. You can see the retrace lines if you maximize the brightness. The scanning raster is a troubleshooting tool that will be described later in this book.

In addition to the scanning raster, a camera must also determine the amount of red, green and blue in the original picture, and the brightness of each square that makes up the picture. When this information is transmitted to the receiver, the receiver must be able to reproduce the original picture by combining correct proportions of red, green and blue, and by reproducing the brightness of each square that defines the picture. All of this picture information is sent to the receiver in video signals.

Using the controls on the television, it is possible to adjust the brightness and contrast to washout or intensify the color of the picture. Even slight adjustments can cause light areas in a picture to lose detail and the dark areas to appear lighter. These problems can be solved easily using the television's controls. Other problems with the color and brightness are not so easy to identify. Later chapters provide more information about analyz-

CHAPTER 1: TELEVISION BASICS

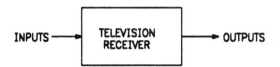

Figure 1-3. A simple block diagram that shows a black box containing circuit information.

ing the color and brightness of the picture produced by the television when troubleshooting picture problems.

Sound Basics

Television sound (audio) is transmitted as FM signals because FM signals are less noisy than AM signals. At the television station, the FM audio signals are combined with the video transmission. Chapter 6 describes how the sound system works for monaural (mono) as well as stereo sound, as well as how to use the television's sound to locate troubles in a receiver.

Using a Block Diagram

Television manuals and repair documentation usually include block diagrams. These diagrams break down the entire process of receiving signals, converting signals, and producing picture and sound in a chart format that makes it easy to follow the sequence of the processes called stages. Thus, block diagrams can be a useful troubleshooting tool, even though they do not replace the more complex schematics that we will discuss later in the book.

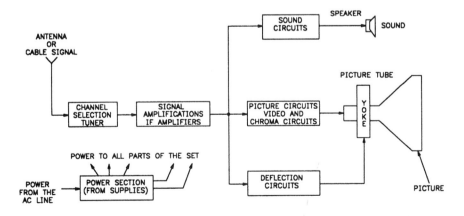

Figure 1-4. A functional block diagram showing the basic steps in the process of converting a signal into televised information.

Block diagrams can show high-level views of a process or complex circuit information. The simplest version of a block diagram is a black box diagram, shown in *Figure 1-3,* in which there is an input and an output, but the process of receiving the signal, converting the signal, and producing picture and sound appears as a black box providing no details.

The block diagram shown in *Figure 1-4* is somewhat more complex. It shows that the television is receiving the appropriate power so that it can complete the required processes. The diagram also shows the basic steps in the process of receiving the signal, converting the signal, and producing the picture and sound, as well as the sequence in which the process takes place.

The antenna receives the signal from the television station. When the television's tuner selects the channel assigned to the television station's transmitting frequency, the television's intermediate-frequency (IF) amplifiers increase the selected signal. Then, the signal is split into the three parts: audio signal, video signal and sync signal. These three signals are processed simultaneously.

Figure 1-5 is a more typical block diagram. This diagram shows all of the processes that take place in the television, from receiving the signal from the television station to producing the video and audio. This diagram makes it easy to trace the signal through each stage of the process.

The following sections describe each stage in the block diagram. Refer to the block diagram in *Figure 1-5* while reading through the sections.

Receiving the Signal

Even though the antenna is separate from the television and is often treated as an independent stage, electrically it is really part of the television's tuner. The antenna receives the RF (radio frequency) signal from the television station and passes the signal to the tuner. All antennas do not receive all television channels. Some antennas work only with VHF channels (channels 2 through 13), and others work only with UHF channels (channels 14 through 83). There also are antennas that receive all channels, as well as some that are designed for only one channel.

Cable companies access signals through a satellite dish. Also, many people now access the television through satellites directly. The signal received from the satellite, whether at a cable company or directly, goes to an ampli-

CHAPTER 1: TELEVISION BASICS

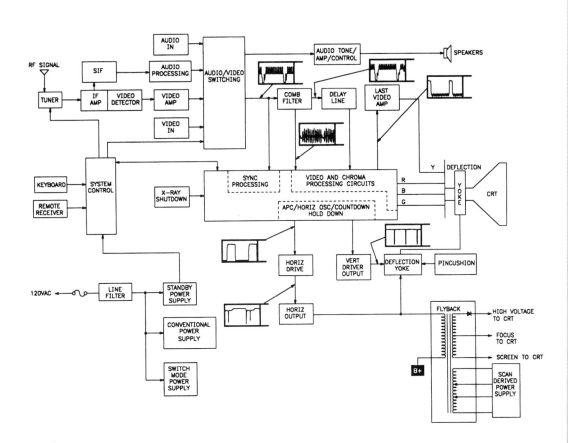

Figure 1-5. A typical block diagram, tracing the signal through each stage of video and audio production process.

fier that lowers the frequency to the television frequency range. Then the signals are passed through a cable to the television. Cable access signals are typically stronger and cleaner than antenna signals.

Each television station transmits at a different frequency. The unit of measurement for a television frequency is megahertz (MHz). The frequencies for the VHF channels are 54 MHz through 216 MHz. The frequencies for the UHF channels are 470 MHz through 890 MHz. The tuner's job is to select a signal at an assigned frequency (channel) from all of the available signals, both weak and strong, then process the signal so that it can be used by the IF (intermediate frequency) amplifiers. IF amplifiers are described in the next section, *Increasing the Signal's Strength*.

After an RF signal is processed, the output is the IF frequency that contains the composite video, audio and sync signals. *Figure 1-6* shows the waveform of the IF signal.

1. 39.75 MHz—Adjacent channel's video carrier.
2. 41.25 MHz—Audio carrier.
3. 41.67 to 42.67 MHz—Color carrier.
4. 45.75 MHz—Video carrier.
5. 47.25 MHz—Adjacent channel's audio carrier.

The IF signal is sent to the IF amplifiers where its strength is adjusted. If all signals were of equal strength, this adjustment would not be necessary. However, all signals are not of equal strength. So, the adjustment is made by a control signal called the automatic gain control (AGC). The AGC is described in the next section.

Increasing the Signal's Strength

Under normal conditions, the signal from the tuner is not strong enough to operate the picture and sound. Therefore, the signal strength has to be increased. The IF amplifiers, shown in *Figure 1-7*, amplify the signal strength.

The AGC, shown in *Figure 1-7*, monitors the average strength of the sync signals. The AGC adjusts the signal's strength by increasing or decreasing a control voltage to the tuner and the IF amplifiers. If the voltage level of the sync signals is too low, the AGC increases the control voltage to the tuner and the IF amplifiers. If the voltage level of the sync signals is too high, the AGC decreases the control voltage to the tuner and the IF amplifiers.

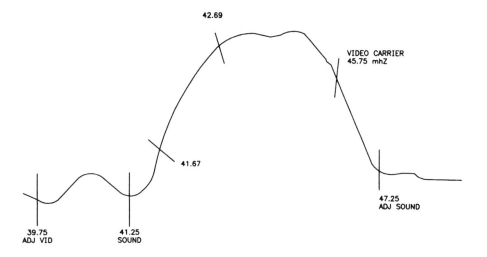

Figure 1-6. *The waveform of an IF signal.*

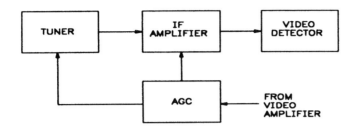

Figure 1-7. Block diagram of an IF amplifier and AGC.

Separating Video and Audio

After the signal is amplified sufficiently in the IF amplifiers, the signal is passed to the video detector, shown in *Figure 1-8*. The video detector is in the integrated circuit (IC) that is part of the video IF amplifier and converts the amplified signal from the IF amplifiers and separates the signal into two types:
1. The composite video, which is an AM signal (30 Hz to 4.2 MHz) containing the video and the sync signals.
2. The sound IF, which is an FM signal (4.5 MHz).

Then, the composite video signal and the sound IF signal are sent to the video amplifiers.

Processing the Video

The video amplifiers perform several tasks. First, the video amplifiers increase the signal from the video detector. Then, the sound IF signal and sync signals are separated from the picture information. The sound IF signal is sent to the sound IF amplifier. The sync signals are sent to the sync separator.

The video signal is amplified more, then sent to the video processing circuit. The contrast control, located in the video processing circuit, controls the amount

Note: The video information is modulated as an AM signal. The AM signal is used to set up exact bandwidths.

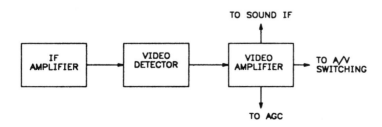

Figure 1-8. Block diagram of a video detector and video amplifier.

of video sent to the CRT, like the volume control adjusts the audio level sent to the speakers. In color televisions, the color signals are separated at this point and sent to the luminance and chroma processing stage. Color processing is discussed later in Chapter 5.

Processing the Audio

The audio processing system, shown in *Figure 1-9*, receives an audio signal from the video amplifier and processes it so that the speakers can produce the sound that accompanies the picture.

The sound IF amplifier increases the FM audio signal before it is sent to the sound detector. Before the audio signal can be sent to the speaker, it must be separated from the sound carrier by the sound detector. The output from the sound detector is an electrical audio signal that must be amplified again by the audio amplifier before it can be heard through the speaker. After the sound completes its trip through the sound processing system, the speaker converts the electrical audio signal to audible sound.

The volume control and tone control are in the audio amplifier stage.

Processing the Sync Signals

Remember that the sync signals are sent by the television station along with the audio and video signals, and ensure that the picture that appears on the screen is synchronized with the picture that was transmitted. Loss of sync signals can cause many problems, from no video to partial picture loss and tearing.

The sync amplifier increases the sync to the required level, and makes sure the correct video timing and placement information is present. The sync separator separates the sync signals into horizontal sync signals and vertical sync signals. The horizontal sync signals control the starting time of the left-to-right lines on the screen. The vertical sync signals control the starting time of

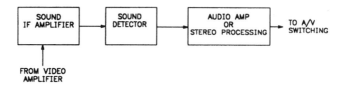

Figure 1-9. Block diagram of a sound IF amplifier, sound detector, and audio amplifier.

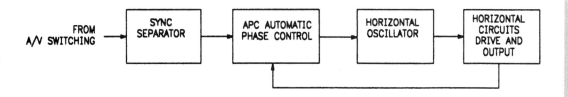

Figure 1-10. A horizontal oscillator and output, APC, and a high-voltage power supply.

the picture from the top-left corner of the screen. As mentioned previously, problems with the sync signal processing can cause vertical rolling or horizontal tearing.

Horizontal Sync Signals

The horizontal sync signals are sent to the automatic phase control (APC), shown in *Figure 1-10*. The rate at which the horizontal output operates determines the actual rate at which the picture is scanned to the screen. The picture scan rate must be synchronized with the rate at which the picture is transmitted. To make sure the horizontal sync signal is synchronized with the sync signals, the output is compared to the sync signals by the APC.

After the comparison is made, any adjustments to the horizontal signal are made by the horizontal oscillator. If the rate is too slow, the APC causes the oscillator to speed up, which produces a faster horizontal sync signal rate. If the rate is too fast, the APC causes the oscillator to slow down, which produces a slower horizontal signal rate. In this way, the horizontal oscillator produces a horizontal scan signal that is correctly synchronized with the sync signals. The horizontal hold control is in the horizontal oscillator. Next, the horizontal signal is sent to the horizontal output stage.

The horizontal output stage is a powerful signal amplifier, which is turned on and off by the horizontal oscillator. The main job of the horizontal output stage is to provide adequate horizontal scan power to the deflection yoke, an electrical assembly attached to the neck of the CRT which controls the horizontal and vertical scan that produces the picture. The output signal causes the electron beam in the CRT to scan horizontally at a rate that is synchronized with the sync signal—15,750 times every second—to produce one frame of the video. A considerable amount of power is needed to move the electron gun so rapidly. The television's horizontal output stage, also shown in *Figure 1-10*, provides the required signal to generate the high voltage needed by the CRT.

Vertical Sync Signals

The vertical sync signals are sent to the vertical oscillator section. Then, after amplification, the signal is sent to the vertical deflection yoke. The output signal causes the electron beam in the CRT to scan vertically at a rate that is synchronized with the sync signals. The vertical hold control, and the picture height and linearity controls are in the television's vertical scan stage.

Producing the Picture

The CRT, shown in *Figure 1-11*, receives input from several sources in the television. It receives the voltage necessary to operate the electron gun from the high-voltage power supply which is in the flyback transformer stage. The video signals come from the video amplifiers. The horizontal and vertical scan, processed from the sync signals, provide the timing and location information for the video. The video signal provides the chrominance (color) for color televisions, and luminance (intensity) for both color and black-and-white televisions.

From these inputs, the CRT reproduces the picture transmitted by the television station. If all parts of the television work properly, the reproduced picture is synchronized with the transmitted picture, and the reproduced picture is high quality, without flickering or distortions. However, all parts of the television do not always work properly, and a low quality reproduced picture or no picture can result. The remainder of this book describes troubleshooting methods that can help you locate and repair problems that occur.

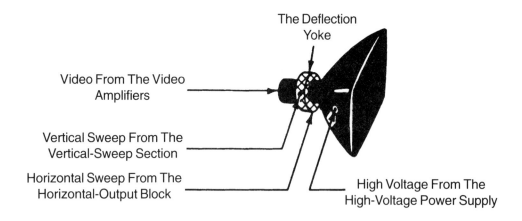

Figure 1-11. A CRT diagram.

Quiz

1. Define the following terms:
   a. Antenna
   b. Audio
   c. Automatic phase control (APC)
   d. Automatic gain control (AGC)
   e. Deflection yoke
   f. Electron gun
   g. Frame
   h. Horizontal oscillator
   i. Horizontal sync signals
   j. IF amplifier
   k. Interlacing
   l. Sync signals
   m. Tuner
   n. Vertical sync signals
   o. Video
2. How many horizontal lines does it take to produce one frame?
3. What is the purpose of the sync separator?
4. How is flickering reduced in a picture?
5. What are the two functions of the video amplifier?
6. What are the three types of signals transmitted over a carrier signal from the television station?
7. Why are FM signals used to transmit audio information?
8. Why is a block diagram a useful troubleshooting tool?

Key

1.a. A device that collects the composite RF signal transmitted from the television station.
1.b. Any sound, mechanical or electrical, that can be heard. Audio is normally 20 Hz to 20,000 Hz.
1.c. Holding the horizontal oscillator in step (phase) with the horizontal sync signal.
1.d. Regulating the receiver's overall amplification (gain) to produce a constant output from variable input.
1.e. An electrical assembly attached to the neck of the CRT which controls the horizontal and vertical scan that produces the picture.
1.f. An assembly in the CRT that emits a small electron beam.
1.g. One television picture in which there are 525 horizontal lines.
1.h. An oscillator that makes adjustments to the horizontal sync signal.
1.i. Right-to-left instructions that control the horizontal scanning on the screen.
1.j. Amplifier that increases the signal strength from the tuner.
1.k. A method of scanning the 525 horizontal lines in two successive fields of 262+ lines to reduce picture flicker. After the first field is scanned, the lines from the second field are scanned between the lines of the first field.
1.l. A series of signals or timing signals that synchronize the picture scanned on the screen with the picture transmitted from the television station.
1.m. The circuit that selects the channel assigned to the television station's transmitting frequency.
1.n. Bottom-to-top instructions that control vertical scanning on the screen.
1.o. Picture signals that are used to produce the picture on the screen.
2. 525.
3. Circuit that separates the sync signals into horizontal sync signals and vertical sync signals.
4. The 525 horizontal scan lines are interlaced in two groups of 262+ fields.
5. Amplifies the signal from the video detector and separates the sound IF signal and the sync signals from the picture information.
6. Composite video, audio and sync signals.
7. FM signals are less noisy than AM signals.
8. They show high-level views of a process or complex circuit information.

Chapter 2
WORKING SAFELY

Chapter 2
Working Safely

The importance of safety can never be stressed enough. The leading manufacturers include safety information and specific warnings with the televisions when they ship them from the factory. However, most of this information is designed as general operating guidelines for consumer consumption, not for technical troubleshooting and repair.

This chapter discusses specific safety guidelines for technicians who work on integrated circuits, power supplies and high-voltage circuits. If a technician does not follow the appropriate safety guidelines, it can be hazardous to the technician and to the consumer whose television the technician is repairing.

Only qualified service technicians who are familiar with safety checks and guidelines should perform service work. Before replacing parts, disconnect the power source to protect yourself and electrostatically sensitive parts. Do not attempt to modify any circuit unless directed to do so by the television's manufacturer.

General Guidelines to Follow Before Returning the Repaired Receiver

Perform a final safety check before returning a receiver to the customer:
1. Check the repaired areas for poorly soldered connections. Make sure that solder joints are ball-shaped, shiny and smooth, not sharp. If the solder is grainy and dull, there is an air bubble in the center of the solder. This air bubble will cause a bad solder joint.
2. Check the entire circuit board for solder splashes.
3. Check the inner board wiring for wires that are pinched or frayed, or that touch any high-wattage resistors.
4. Check that all control knobs, shields, covers, grounds and mounting hardware have been replaced.
5. Replace all insulators and restore properly dressed leads.
6. Check all leads for strain or tension, especially those near any metal parts.
7. Make sure that all insulation and isolation materials are replaced.
8. Remove any loose or foreign particles from the circuit board and the cabinet.

CHAPTER 2: WORKING SAFELY

9. Make sure that all hardware items have been reinstalled according to the factory specifications.
10. Make sure that all replacement parts meet safety requirements and factory specifications for the specific television brand and model.

Testing for Cold Leakage Current in Receivers with Isolated Ground

To test for cold leakage current:
1. Unplug the television's AC power cord.
2. Connect a jumper across the power cord's plug prongs.
3. Turn on the power switch (if applicable).
4. Using an ohmmeter, measure the resistance between the jumped AC plug and any exposed metal cabinet parts, such as the antenna terminals, control shafts or handle brackets.
5. Read the ohmmeter. Exposed metal parts with a return path should measure between 1 megohms and 5.2 megohms. Exposed metal parts without a return path must measure infinity.

Testing for Hot Leakage Current

To test for hot leakage current:
1. Plug the television's AC power cord into a 120V AC outlet.
2. Using an AC voltmeter or a volt-ohmmeter with a sensitivity of at least 5000 ohms per volt, measure the voltage across the resistor, as shown in *Figure 2-1*. Check all exposed metal parts and measure the voltage at each point.
3. Using a 1500 ohm, 10W resistor in parallel with a .15 mF capacitor, connect one clip to any exposed metal part on the receiver and the other clip to a good earth ground.

Note: Do not use an isolation transformer when testing for hot leakage current.

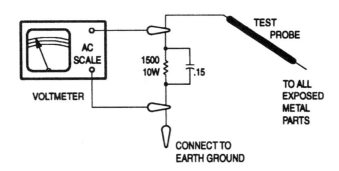

Figure 2-1. A voltmeter, measuring voltage across a resistor.

4. Note the meter's AC voltage drop across the resistor. Voltage measurements should not exceed .75V AC, 500 mA. Any value that exceeds this limit constitutes a potential shock hazard and must be corrected.
5. If the AC plug is not polarized, reverse the AC plug and repeat Steps 4 and 5.

Avoiding Electrical Shocks When Servicing High-Voltage Circuits and CRTs

Use extreme caution when servicing high-voltage circuits:
1. Perform the tests for cold and hot current leakage.
2. Always use tools that have been approved for use on electrical circuits. Also use rubber pads on the work surface and on the floor beneath the work surface.
3. Always wear shatterproof goggles when handling the CRT to protect the eyes in case of implosion.
4. Read thoroughly any instructions and procedures that address high voltage.
5. Make sure that all test equipment is in good repair, is calibrated properly and works correctly. Keep an accurate high-voltage meter available at all times, and check the meter's calibration frequently.
6. When servicing the receiver, use an isolation transformer between the line cord and the power receptacle.
7. Discharge the CRT before removing it from the chassis. To discharge static high voltage, connect a 10 ohm, 15W resistor in series with a test lead between the chassis ground and the CRT's anode lead. Do not lift the CRT by the neck.
8. When servicing a receiver, check the voltage at various brightness levels to make sure the receiver is regulating properly.
9. Use plastic knobs and bushings to isolate metal parts, and replace missing or broken insulators.
10. Make sure that the power cords and all leads are not frayed and do not have exposed wires. Also, make sure the power cord is unplugged before you touch components, such as the flyback transformer or horizontal output transistor.
11. Make sure that all antenna leads are isolated from the chassis using capacitors and resistors.
12. Replace any part that is deteriorated.
13. Discharge the filter capacitors before checking voltages in the power circuits.
14. Connect the lead of any high-voltage probe to a good earth ground.

CHAPTER 2: WORKING SAFELY

15. Avoid voltage arcing (corona) that might result from pointed solder joints or frayed wires. Voltage arcing can destroy the solid-state circuits in the receiver, damage the CRT, and cause fires.
16. Avoid static charge buildup by using a grounding wrist strap or a static discharge rug.

Avoiding X-Ray Radiation and High-Voltage Limits

X-ray radiation is produced by excessively high voltage. In solid-state receivers and monitors, the CRT is the only potential source of X-rays. In some receivers, many electrical and mechanical components have safety-related characteristics which are not detectable by visual inspection. These components, such as the high-voltage rectifier and the shunt regulator tube in older televisions, are identified by a number on the schematic and on the parts list:

1. Read thoroughly any instructions and procedures that address X-ray radiation.
2. Use a correctly calibrated high-voltage meter to test for excessive voltage. An example is shown in *Figure 2-2*.
3. Keep high-voltage values at the rated value and no higher. Do not depend on protection circuits to keep voltage at the rated value. Excessively high voltage may cause X-ray radiation or failure of the associated components.
4. Discharge the CRT before working on it.
5. When troubleshooting a receiver with excessively high voltage, avoid close contact with the CRT and do not operate the receiver longer than is necessary.
6. To locate the cause of excessively high voltage, use a variable AC transformer, shown in *Figure 2-3*, to regulate voltage.
7. Use only equivalent replacement parts when replacing components that may be a source of X-ray radiation.

Figure 2-2. A high-voltage meter.

Figure 2-3. A type of variable AC transformer.

Avoiding Fire Hazards

Follow these rules when working on the television to reduce the chances of creating a fire hazard:
1. Do not let flammable materials come in contact with the CRT, power resistors or output transistors.
2. Prevent voltage arcing and static charge buildup.
3. When replacing a fuse or circuit breaker, use an exact match as a replacement.
4. Never jump fuses with wire or other conducting material.

Working with a CRT

When working with a CRT, it is possible to receive a severe shock, a burn from extremely hot connections, or an injury from an implosion—a rapid inrush of air when the vacuum is broken. When an implosion occurs, shards of glass can cause injuries. Therefore, follow these guidelines when working with CRTs:

1. Before working with a CRT, slowly discharge the anode connection using a 10 megohm, 15W or greater resistor, and connecting it to a good ground. Do not short the anode connection to ground instead of discharging the anode connection.
2. Check the CRT carefully for broken seals, cracks, scratches or any other flaws in the surface.
3. Always wear safety glasses and gloves when handling a CRT, and do not handle a CRT when other people are nearby.
4. Safely dispose of defective CRTs. Some CRT manufacturers offer a recycling program for disposing of old or defective CRTs, many times at the manufacturer's expense. For information, call the manufacturer of the CRT you are replacing and inquire whether they offer a CRT recycling program.
5. Always replace a CRT with one that is factory authorized for the model and brand of television for which the replacement is intended.
6. Do not overtighten the mounting hardware.
7. Make sure the CRT and any connected parts are properly grounded.

Note: *There are now laws on the disposal of CRTs due to high lead content in the glass. Be sure to dispose of CRTs in an approved manner.*

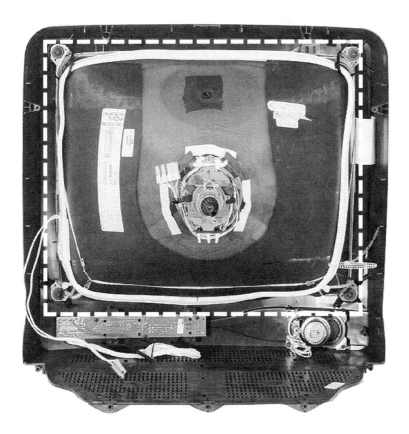

Figure 2-4. A degaussing coil.

8. For color televisions, make sure the degaussing coil (shown in *Figure 2-4*) is not punctured or pinched. With the degaussing coil unplugged, check each lead using an ohmmeter. If the coil is good, the ohmmeter should read very low resistance.

General PHOTOFACT Safety Symbols

When you work with PHOTOFACT television schematics, be aware of the following symbols:

- **#** : For safety, use only equivalent replacement part; see parts list.
- **✳** : Circuitry not used in some versions of this particular TV.
- **---** : Circuitry used in some versions of this particular TV.
- **⏚** : Ground.
- **⏛** : Chassis ground.
- **▽** : Common tie point.
- **△** : Taken from common tie point.
- **3** Schematic CircuiTrace® : Voltage source tie point.
- **A —** : Cabling; heavy lines reduce the use of multiple lines.

General PHOTOFACT Safety Precautions

The following safety tips are commonly given in PHOTOFACT to aid the repair technician and prevent safety problems:

Before replacing parts, disconnect power source to protect electrostatically sensitive parts. Do not attempt to modify any circuit unless so recommended by the manufacturer. When servicing the receiver, use an isolation transformer between the line cord and power receptacle.

Use extreme caution when servicing the high voltage circuits. To discharge static high voltage, connect a 10K ohms resistor in series with a test lead between the receiver and CRT anode lead. DO NOT lift the CRT by the neck. Always wear shatterproof goggles when handling the CRT to protect eyes in case of implosion.

Be aware of the instructions and procedures covering X-ray radiation. In solid-state receivers and monitors, the CRT is the only potential source of X-rays. Keep an accurate high voltage meter available at all times. Check meter calibration periodically. Whenever servicing a receiver, check the high voltage at various brightness levels to be sure it is regulating properly. Keep high voltage at rated value, NO HIGHER. Excessive high voltage may cause

CHAPTER 2: WORKING SAFELY

X-ray radiation or failure of associated components. DO NOT depend on protection circuits to keep voltage at rated value. When troubleshooting a receiver with excessive high voltage, avoid close contact with the CRT. DO NOT operate the receiver longer than necessary. To locate the cause of excessive high voltage, use a variable AC transformer to regulate voltage. In present receivers, many electrical and mechanical components have safety-related characteristics which are not detectable by visual inspection. Such components are identified by a # on both the schematic and the parts list. For SAFETY, use only equivalent replacement parts when replacing these components.

For cold leakage checks for receivers with isolated ground, unplug the AC cord, connect a jumper across the plug prongs, and turn the power switch on (if applicable). Use an ohmmeter to measure the resistance between the jumped AC plug and any exposed metal cabinet parts such an antenna screw heads, control shafts, or handle brackets. Exposed metal parts with a return path should measure between 1M ohms and 5.2 ohms. Parts without a return path must measure infinity.

For a hot leakage current check, plug the AC cord directly into an AC outlet. DO NOT use an isolation transformer. Use a 1500 ohms, 10W resistor in parallel with a .15µF capacitor to connect between any exposed metal parts on the receiver and a good earth ground. Use an AC voltmeter with at least 5000 ohms per volt sensitivity to measure the voltage across the resistor. Check all exposed metal parts and measure voltage at each end. Voltage measurements should not exceed .75 VAC, 500µA. Any value exceeding this limit constitutes a potential shock hazard and must be corrected. If the AC plug is not polarized, reverse the AC plug and repeat exposed metal part voltage measurement at each point.

Perform a final SAFETY CHECK before returning receiver to customer. Check repaired area for poorly soldered connections, and check entire circuit board for solder splashes. Check inner board wiring for pinched wires or wires contacting any high wattage resistors. Check that all control knobs, shields, covers, grounds, and mounting hardware have been replaced. Be sure to replace all insulators and restore proper lead dress.

Quiz

1. Why is it important to discharge a CRT before working on it?
2. What are two sources of arcing?
3. What is the main source of X-ray radiation in a television?
4. What is an implosion?

Key

1. So that the technician will not be shocked.
2. Pointed solder joints or frayed wires.
3. Excessively high-voltage power supplies.
4. A rapid inrush of air when the vacuum in a CRT is broken.

Chapter 3
THE BASICS OF TROUBLESHOOTING

Chapter 3
The Basics of Troubleshooting

There are many techniques you can use to troubleshoot problems with televisions. Most of these techniques require specialized electronic equipment. This chapter describes the basic techniques you can use to test the electronic components that make up a television. Later chapters will describe specific techniques for specific problems.

Basic Equipment for Troubleshooting Televisions

Examples of the basic pieces of equipment that help in troubleshooting problems with televisions include:
1. Oscilloscope.
2. Digital multimeter (DMM) or a volt-ohmmeter (VOM).
3. CRT tester.
4. CRT test jig.
5. Color bar generator (NTSC).
6. Degaussing coil.
7. High-voltage probe.

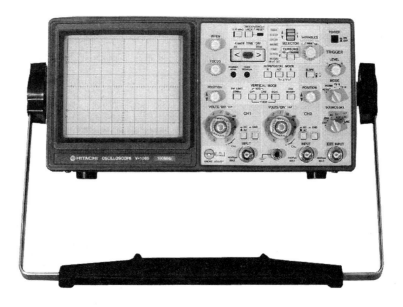

Figure 3-1. *An oscilloscope.*

CHAPTER 3: THE BASICS OF TROUBLESHOOTING

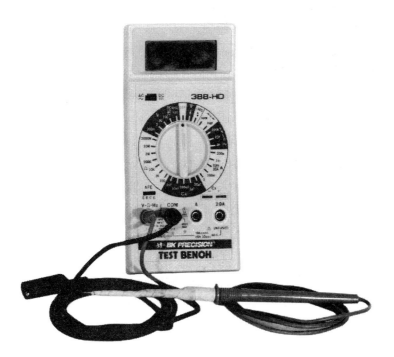

Figure 3-2. A digital multimeter (DMM).

An oscilloscope, shown in *Figure 3-1*, is like a voltmeter that measures changes to AC waveforms and DC voltages and displays the changes on a screen. However, unlike a voltmeter, the oscilloscope can measure and display very rapid changes to voltages like those found in televisions. Also, the waveforms are traced to the oscilloscope's screen, letting you see the shape and size of the waveform.

A digital multimeter (DMM), shown in *Figure 3-2*, and a volt-ohmmeter (VOM), can be useful tools for measuring current, voltage and resistance. They also can be used to check capacitors, diodes, and transistors. DMMs can make finer measurements than volt-ohmmeters, and multimeters can be analog or digital. When choosing a DMM, make sure the DMM can read the switching circuits (low ohms function).

A CRT tester, like the one shown in *Figure 3-3*, can be used to test a picture tube that is not working correctly. A CRT tester is portable and can be used to test a CRT in someone's home, if necessary. However, the CRT test jig shown in *Figure 3-3* is a better diagnostic tool on the bench because it can replace the picture tube and be used to quickly determine if a picture problem is in the CRT or the receiver.

Figure 3-3. A CRT test jig.

A color bar generator, shown in *Figure 3-4*, can produce color bar, dot and crosshatched patterns that help when working with color setup procedures because the signal source has known properties. The color bar generator must meet NTSC standards.

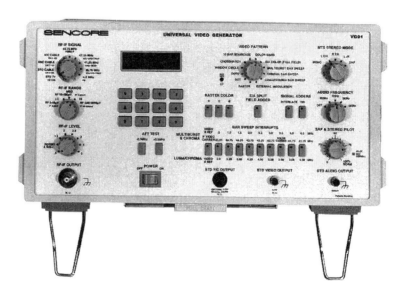

Figure 3-4. A color bar generator.

Figure 3-5. A manual degaussing coil.

The manual degaussing coil, shown in *Figure 3-5*, demagnetizes the CRT and improves the picture's quality. When color CRTs become magnetized, picture quality is greatly reduced.

A high-voltage probe, like the one in *Figure 3-6*, is used to detect the high voltage used by the television. It can be used as a diagnostic tool to make sure the CRT has the correct high voltage when there is no picture present or when the picture quality is poor or varies. It also can be used to detect excessively high voltage which causes X-ray radiation.

General Techniques for Servicing Televisions

Before you start, make sure you have the proper equipment, and that the equipment is properly calibrated and in good working order. Also, make sure you have the proper documentation and schematics for the brand of television you are repairing. Then, follow these general guidelines for locating the source of the problem.

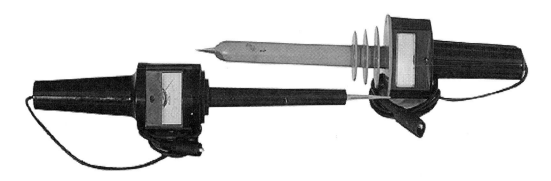

Figure 3-6. A high-voltage probe.

Take your time, but don't spend hours tracking a problem with no results. Sometimes you might have to contact the television's manufacturer for help:

1. Ask the customer for a description of the problem. It might be helpful to ask what happened prior to the appearance of the problem.
2. Observe the video product and listen to the audio product. Look for the apparent problems, but do not overlook the subtle distortions. Look at the chassis and see if you can locate any burned areas. Also, look for any cracked solder connections or burned resistors.
3. Listen for any unusual sounds, such as a ticking sound when the flyback transformer receives excessively high voltage. You can hear static and the cracking noise arcing makes. You also can hear the hums that components can make when they are not working properly.
4. Notice any unusual smells that the television produces. For example, is there an odor of overheated components or ozone which is typically from a burned resistor or excessive high voltage? Also, feel the outside of the television to see if there are any excessively hot areas. The vertical and horizontal stages, and the power supply can produce excessive heat.
5. Check the fuses. Perform a visual check and measure the resistance of the fuse.
6. Look at the schematics for the specific model and brand of television to see if you can locate any obvious problem source.
7. Test the controls and observe the effects of the adjustments. While doing so, use the schematic to see if you can isolate the circuit that could be the source of the problem.
8. When you isolate the circuit(s), determine their inputs and outputs. Also, locate the power source for each of the suspected problem circuits. Check the circuits that precede and follow the suspected circuit(s). Again, the schematic is very helpful in this step.
9. Use the oscilloscope to make sure the input and output signals are the appropriate strength for the suspected circuit(s) and its supporting circuits or components. A loss of signal can be caused by an open series transistor or a shorted parallel component. When a transistor is open, the collector voltage is very high, even when there is no voltage applied to the emitter terminal.
10. Before replacing the suspected component, check the entire circuit to which it belongs. Make sure that there are no faulty components or connections around the component you are replacing.
11. Before returning the television to its owner, perform a complete safety check on the television as outlined in the section *General Guidelines to Follow Before Returning the Repaired Receiver,* in Chapter 2. Remember to ground yourself before performing the safety checks, especially if checking the ICs or any solid-state circuits.

Troubleshooting Methods

The main electronic troubleshooting methods are:
1. Using signal injection and signal tracing.
2. Analyzing circuit voltage.
3. Measuring resistance.
4. Substituting parts.

Signal Injection and Signal Tracing

Signal injection is inputting a signal that is produced outside of the television into the circuit that you have identified as the possible source of the problem. Normally, you would start injecting the signal at a component's output point.

For example, if you inject a video signal into the output of the first video amplifier and the CRT produces a picture, the circuits are working properly between point where the signal was injected and the CRT. Then, you would move to the input of the first video amplifier. Inject a signal into the input. If the CRT does not produce a picture, then the video amplifier is the source of the problem.

Signal tracing is the opposite of signal injection. Signal tracing tracks a signal through a circuit to make sure the signal goes from the input point to the output point properly. For example, using an oscilloscope, check the waveform of the input signal to the video amplifier. If the input is correct, check the signal output from the video amplifier. If the signal is not correct, the video amplifier is the source of the problem.

Although signal tracing can be used, signal injection is most used for low-level, high-frequency circuits in tuners and IF circuits. Likewise, signal injection is less effective than signal tracing when testing the audio amplifier. This is because when you are tracking the source of audio distortion, injected signals many times do not sound as distorted as the transmitted signal. In this case, tracing the signal until you locate the source of distortion is better.

When troubleshooting using signal tracing, use a schematic with a picture of the correct waveform in order to compare the waveform the oscilloscope is producing to the waveform that is expected. However, if a schematic is not available, follow these guidelines to help determine what a circuit is and its purpose:
1. Use a data or substitution book to locate the transistor. Find out what the transistor is made of—silicon, germanium, NPN (Negative-Positive-Negative) or PNP (Positive-Negative-Positive). Also, find out

what the transistor is used for and about its basing. The basing is important because a replacement transistor does not always have the same basing as the component it is replacing. If there is no basing available, you must perform the following test to determine the correct element to use. *Figure 3-7* will help familiarize you with the structure of a transistor.

2. Use an ohmmeter to see if any element, such as the base, collector or emitter on a transistor, is grounded.

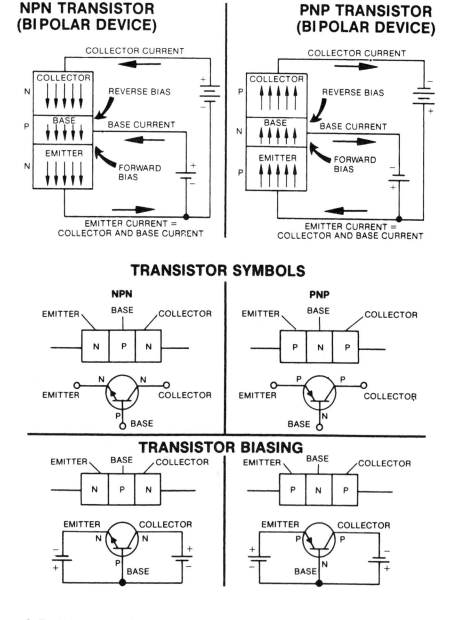

Figure 3-7. *Diagrams showing the structure of two transistors, NPN and PNP.*
(Courtesy of Van Valkenburgh, Nooger & Neville, Inc.)

3. Try to determine the base by following the signal of the preceding stage. If the transistor is being used as an amplifier, the output at the collectors will be of higher amplitude, and inverted. If the transistor is being used as an emitter-follower (common collector), the output signal on the emitter will be approximately the same as the base signal, and not inverted. On many receivers, the video stage is made up of common collector amplifiers—the output signal from the emitter is the same amplitude and the same polarity as the input signal.
4. Using signal tracing, identify which is the output lead by determining which of the transistor's leads is coupled to the next stage in the process. Stages are connected by capacitors, resistors, and coils or transformers. Also, stages are many times arranged from left to right on the chassis. This means that the initial input is on the left side on the chassis. Then, the next stage follows to the right, and so on. Schematics follow the left to right pattern of transistors on the chassis.
5. Normally, by measuring voltages from ground, the base and emitter are easy to locate. If you know the type of transistor, such as a NPN, the most negative voltage is the emitter. The collector is reverse biased or positive. The base and the emitter have a voltage difference of 0.03V for germanium or 0.07V for silicon. However, this method of identifying what the transistor is made of is effective only if the amplifier is working almost normally and is used as a class A amplifier. Class A amplifiers run all of the time. Class B amplifiers turn off after a time. Class C amplifiers act as a switch for other amplifiers. If the circuit is operating as a switch (class C), these values are not reliable.

When troubleshooting, paying close attention to details is very important. For example, if the output signal at the collector is the same amplitude and polarity as the base, and the circuit is a common emitter circuit, the problem could be a shorted transistor which allows the base signal to be applied directly to the collector lead. A circuit that is operating correctly will have output at the emitter, and the output will be the same amplitude and polarity as the input.

When you locate the defective stage, test the transistors in that stage first. Component failures usually happen in the following order; therefore, it might save time to check the components in this order:
1. Active components (such as power transistors, large current resistors or small electrolytic capacitors) amplify, switch or control a function. High voltages and excessive heat can cause the active power components to fail. Check transistors first because they produce heat, and are susceptible to power surges and spikes.

2. Passive components are resistors, low-wattage resistors, coils and low-voltage capacitors. Check the resistors first because they produce heat and can deteriorate over time. Then, check the capacitors because they can be damaged by the heat around them and by power surges. Also, a capacitor's insulation can deteriorate.

Analyzing Circuit Voltages

Measuring the voltage around an integrated circuit or transistor can many times help you locate a defective component. For example, you can use a DMM with the diode test function to measure the voltage at a transistor. Also, high voltage to the collector might indicate an open transistor or one that is biased to cutoff.

To see whether the transistor is open or biased to cutoff, connect a voltmeter across the transistor from the emitter to the collector. Then, connect a 100K resistor from the collector to the base.

The base must be positive for the transistor to conduct. If the transistor biases on, the voltmeter shows a decrease in the voltage drop across the transistor. This indicates that the transistor is good and can control the current passing through it. *Do not short the base collector junction.* This can damage the transistor.

If the emitter's voltage measurement is too high, voltage drop across the emitter resistor will be too high. This can indicate that too much current is passing through the resistor or that the resistance has increased. If too much current is passing through the resistor, it is conducting too heavily or there is an open bypass capacitor. Therefore, if the capacitor opens, the signal can degenerate and cause a large voltage drop across the resistor. Check the base voltage to see if it is causing the transistor to conduct too heavily. The result can be an increase in forward bias—positive for an NPN transistor and negative for a PNP transistor.

If the emitter's voltage is too low, the transistor might not be conducting correctly or may be turned off because, if the voltage is too low or the polarity is incorrect, the transistor will not conduct. Also, if the emitter's voltage is too low, the bypass capacitor might be shorted or leaky—letting current flow where it is not supposed to. It also might mean that the emitter's resistor might be faulty. If the capacitor is not shorted or leaky, and the resistor is not faulty, the transistor is probably damaged.

CHAPTER 3: THE BASICS OF TROUBLESHOOTING

Measuring Resistance

Sometimes it can be useful to use Ohm's law to troubleshoot a resistor or a resistor array. Measure the resistance in parallel or series, use Ohm's law to check if you have the correct value. Ohm's law states that for any circuit, the electrical current is directly proportional to the voltage and inversely proportional to the resistance.

Before you measure a circuit's resistance, unplug the television. Then, when you measure a circuit's resistance, think about the following:
1. Check the schematic before measuring a circuit's resistance to make sure there are no parallel paths that you could measure simultaneously. Parallel paths decrease the total resistance. Just to make sure you are not picking up stray resistance from other components, gently lift one side of the component, especially diodes. This breaks the circuit and ensures that no stray resistance is being measured. Then, test the component.
2. When measuring a circuit's resistance, having a nonparallel path can cause a component to measure a greater value than its rated value. Therefore, if the ohmmeter's leads are placed directly on each end of the component, there is no way that the component can measure greater resistance than its rated value unless the component has increased in resistance.
3. If a component measures less resistance than its rated value when checked in the circuit but normal when measured out of the circuit, a parallel path is being measured. This parallel path could be a junction of a transistor or a diode, or it could be that a capacitor is conducting when capacitors usually do not conduct. This is one way to identify a leaky capacitor. If you are testing connections for transistors or diodes with a low-power ohm range, you might have to remove the component before testing it or use a test voltage that will not cause the connection to conduct.

Substituting Parts

Sometimes part substitution will help in locating the source of the problem. For example, if the picture is collapsing vertically, usually a capacitor is faulty. Take a capacitor of like or greater value and type, and bridge across the suspected faulty capacitor. If picture quality is restored, replace the faulty capacitor.

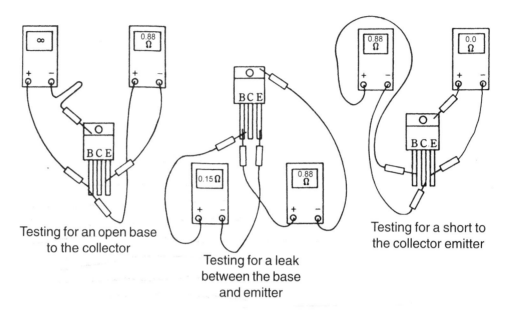

Figure 3-8. Using a digital multimeter to check for shorted, open, or leaky transistors.

Testing Components

The following sections describe the basics of locating defective or faulty components. These topics are described in more detail later in the book.

Locating Defective Transistors

When a transistor is open, it means that a connection is not complete and current is not flowing. A cold soldered joint is an example of an open connection. When you measure leads on the transistor, if any of the leads measure low resistance (0-10 ohms), the lead may be shorted. Be careful. Some transistors might have low resistance diodes and resistors built in. Refer to the schematic from PHOTOFACT for the components and the expected values.

This section discusses testing how well the transistor can control current. To do this, use a transistor tester, or a digital multimeter like the one shown in *Figure 3-8*.

Only use an ohmmeter or a multitester to test for a leaky or open transistor, such as we did in the previous test for locating defective transistors. As they age, transistors can begin to leak current at the emitter-collector junction.

CHAPTER 3: THE BASICS OF TROUBLESHOOTING

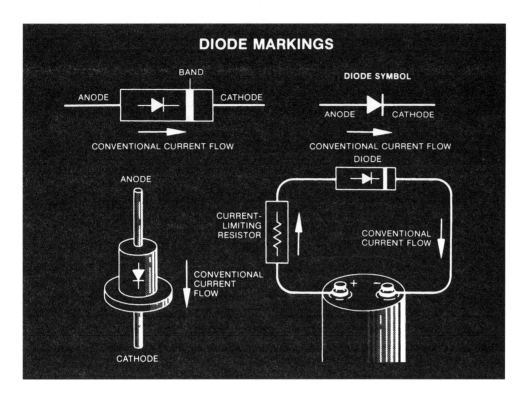

Figure 3-9. A diagram depicting diodes and their markings.
(Courtesy of Van Valkenburgh, Nooger & Neville, Inc.)

Locating Defective Diodes

A diode, shown in *Figure 3-9*, is a device that lets current flow only in one direction. Notice the special markings on the diodes in the figure. The arrows show the conventional current flow. Diodes have polarity; therefore, they must be placed in circuits in the correct direction. Small diodes are used for low-current applications in circuits up to 1 amp of average current and are marked with a band to indicate the cathode end. Large diodes are rated in terms of their peak and average forward current-carrying capabilities and are marked with the diode symbol to indicate their polarity. Do not exceed the rated value of a diode. A transistor is really two diodes, back-to-back.

You can use a multimeter to check whether a diode is open or is leaking current, as we did in the previous section, *Measuring Resistance*.

Locating Defective SCRs

A silicon controlled rectifier (SCR) is basically used as a switch, as shown in *Figure 3-10*. Use a DMM to check the voltage on a SCR.

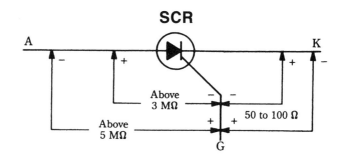

Figure 3-10. Diagram showing the switching action of a silicon-controlled rectifier, or SCR.

Locating Defective ICs

Use a DMM to locate a defective IC. If a component is leaking current, the voltage on the terminals will be low, especially the terminal directly supplied by the low voltage power supply. If you measure a low resistance between the terminal pin and the chassis ground, and there is no external path to ground, you may have a defective IC.

An open connection in the IC can produce an intermittent problem, and intermittent problems are difficult at times to trace. You can apply a cold spray to the IC and then lightly tap it. This sometimes reveals a broken solder joint. However, if you apply a signal to the input and get no output from the IC, and all external components are good, replace the IC.

Surface-Mount Technology (SMT)

Surface-mount technology (SMT) has dramatically changed the television chassis. *Figure 3-11* shows a comparison between SMT and IMCs (insertion-mount components) used in earlier television models. SMT offers smaller and lighter components whose leads are part of the board and that can be soldered directly to the television's wiring. Smaller components are usually located on the bottom of the board and the larger components are usually located on the top of the board. Using double-sided boards, television manufacturers can produce a smaller, lighter, and more compact chassis.

The mounting methods of SMT and IMC components are different. Every solder connection is a possible failure point. Solder connections can weaken or, as with a cold solder joint, not conduct at all. Therefore, when SMT manufacturers reduce the number of solder connections, they reduce the number of potential problems in televisions.

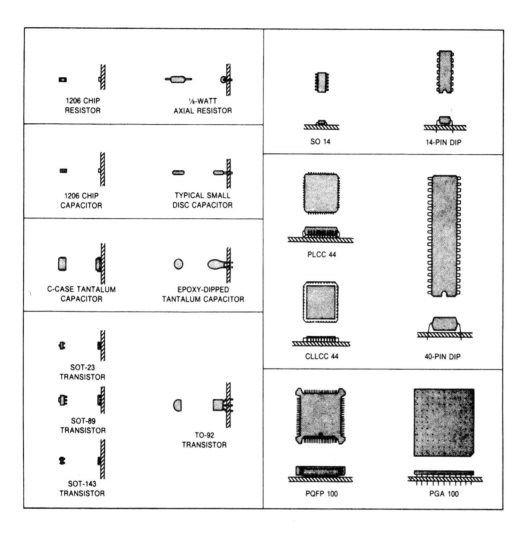

Figure 3-11. A chart comparing SMT and IMC components.
(Courtesy of Jim Hollomon, Jr.)

SMT boards have smaller leads than IMCs. In fact, chip passives (capacitance and resistance arrays) do not have leads. This makes the paths between components shorter, which boosts the performance. Each additional millimeter of lead adds inductance, capacitance and resistance, called parasitics, to the signal path. Reducing the distance between components can improve circuit performance. Also, a component's lead can act as an antenna, receiving radio-frequency signals. By reducing the length of leads, the SMT is provided with some external noise protection.

Some companies are producing "universal replacement" components for SMT boards. However, use care when replacing components; it is best to use an exact factory replacement from the television's manufacturer. The "universal" replacement parts and the exact replacement parts might look the same, but sometimes they are not.

When you remove a surface-mounted component, keep in mind the following suggestions. Always use a soldering iron and a desoldering iron especially designed for SMTs. Never use a soldering gun:
1. Carefully remove the defective component by lightly touching a soldering iron to melt the solder at one end or at each terminal on one side of the component.
2. Using tweezers, gently lift the end from which you removed the solder as you lightly touch the connection on the other end or side of the component with the soldering iron.
3. As the solder melts away from the joint(s), gently lift up and twist the component until it is free.
4. Using the desoldering iron designed for SMTs, remove any solder fragments, and any glue if the component also was glued to the board. Make sure the surface and all leads are clean.

When you replace the component, follow these steps:
1. Preheat the component, except semiconductors and IC processors, using a small, hand-held heater or even a hair dryer. Never preheat a semiconductor or an IC because excess heat can destroy them.
2. Place solder at the contact points.
3. Press the component into the solder, gently but firmly. Be careful when you solder the joints. Excessive heat or heat applied for too long— longer than 3-5 seconds—can destroy the component or lead.
4. Apply the solder directly without rubbing the joint as you apply the solder. Also, be careful not to break the component by applying excessive pressure. These parts are very small and somewhat delicate.

Servicing Solid-State Devices

When you are servicing solid-state components, you can easily destroy them by applying too much voltage or voltage surges, ignoring the polarity, using excessive heat, or exposing them to high-energy magnetic fields. Therefore, test the voltage in a circuit before you replace a transistor or IC. Also, remember to use exact replacements. As mentioned above, "universal" replacements might not be exactly the same as the television manufacturer's replacement. This can cause problems with the operation of the television or possibly damage it. Also, pay close attention to the schematic for the brand and model of television and to the manufacturer's recommendations when replacing parts.

CHAPTER 3: THE BASICS OF TROUBLESHOOTING

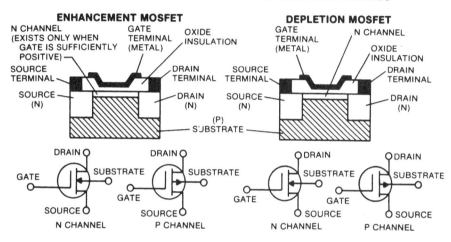

Figure 3-12. MOS and FET devices.
(Courtesy of Van Valkenburgh, Nooger & Neville, Inc.)

When soldering solid-state devices, use a heat sink to draw the soldering iron's heat away from the device.

For MOS (metal oxide semiconductor) devices, ground yourself and the soldering iron to prevent static electricity charges from forming. As with SMTs, use a soldering iron because a soldering gun has a magnetic field around it that can damage MOS or FET (field effect transistor) devices. *Figure 3-12* shows a diagram of a MOS and a FET.

Also, when you cut the leads of MOS and FET devices, use a sharp, scissors-like cutter such as a wire cutter (*Figure 3-13*), and not a pincher cutter. The wire cutters make a clear cut without sending shock waves through the device.

Protect the components from high voltage arcing because it can destroy solid-state devices. Therefore, do not bridge filter capacitors across a faulty capacitor while the television is receiving power. Make sure that the television is well grounded and that all connections are grounded. If you have to remove a ground or place a jumper between two points, turn off and unplug the television first. Also, make sure that jumper wires and the probe wires are

Figure 3-13. Two types of wire cutters.

in good condition. Bare wires and exposed meter terminals can damage a solid-state component, as well as deliver a shock.

When you are using probes to test components, make sure the points of the probes are sharp. The points may have to penetrate oxidation that has formed on the joint. Also, sharp probe points keep the points from slipping as you are using them. Do not touch the leads directly; instead, touch the leads. When you touch the leads directly, the probe points, though sharp, can slip between the leads and cause a voltage arc.

Digital Circuits

It is important to understand digital circuits and learn how to troubleshoot these circuits. The basics of and theory behind digital circuitry is described in basic electronics books. There are advanced books that describe more advanced concepts of digital circuitry and how to troubleshoot a digital circuit. This book will not cover troubleshooting digital circuits.

Helpful Diagrams

The following diagrams depict information commonly available in PHOTOFACT, and will be useful tools for you.

Parts List

The parts list gives manufacturer part numbers and describes the components you will be working with. (*Figure 3-14.*)

CHAPTER 3: THE BASICS OF TROUBLESHOOTING

COILS & TRANSFORMERS

Item No.	Function/Rating	Mfr. Part No.
DY1	Yoke 90° Horiz 1.22mH Vert 15.4mH	4835 150 17102
FB401, 02	Ferrite Bead	4835 526 17002
FB431, 32	Ferrite Bead	4835 526 17002
L1	47µH	4835 157 57763
L200	VCO 45.75MHz	4835 157 57485
L204	AFT	4835 157 57594
L214	Sound Discriminator	4835 157 57113
L230	27µH	4835 157 57119
L301	3.3µH	4835 157 57154
L311	1.2µH	4835 157 67003
L312	12µH	4835 157 57048
L313, 14, 15	4.7µH	4835 157 67011
L357, 58, 59	27µH	4835 157 67019
L365	12µH	4835 157 57048
L386, 87	3.3µH	4835 157 57266
# L400	Line Filter	4835 152 17001
L405	1.8µH	4835 152 27029
L409	100µH	4835 157 57047
L422	2.2µH	4835 157 57752
L423	.68µH	4835 157 57751
L430	.7µH	4835 152 27036
L441	Ferrite Bead	4835 526 17009
L448	10µH	4835 152 27002
L452	2.2µH	4835 157 57752
L459	10µH	4835 157 57093
L460	42µH	4835 157 57063
# L499	Degaussing	4835 157 97072
L502	42µH	4835 157 57673
L503	Horizontal Linearity	4835 150 17101
L602	10µH	4835 150 57039
L605	15µH	4835 157 57756
L606	10µH	4835 150 57006
L610	10µH	4835 150 57039
L620	2.7µH	4835 157 57098
L625	10µH	4835 150 57004
L635	18µH	4835 157 67031
L636, 37, 38	1.2µH	4835 157 67003
L704	160mH	4835 157 57657
L900	15µF	-
# T401	Power	4835 148 87286
# T450	Standby	4835 148 87251
T501	Horizontal Drive	4835 142 47018
# T502 (1)	Horizontal Output	4835 140 67088
V421 (L421)	.7µH	4835 152 27036
V546	5.6µH	4835 152 27038

For SAFETY use only equivalent replacement part.
(1) Focus and screen controls are part of T502.

COILS & TRANSFORMERS continued

Item No.	Function/Rating	Mfr. Part No.
PIP BOARD		
L101	1.2µH	4835 157 67028
L104	6.8µH	4835 157 57061
L105	.47µH	4835 157 57014
L107	6.8µH	4835 157 57061
L201	100µH	4835 157 57818
L202	6.8µH	4835 157 57061
L301	.47µH	4835 150 57045
L302	39µH	4835 157 67021
L303	3.3µH	4835 157 57832
L304	6.8µH	4835 157 57061
L305	10µH	4835 150 57008
L306	12µH	4835 157 57831
L307	5.6µH	4835 157 57833

CABINET PARTS

Item	Mfr. Part No.
Button Assembly	4835 219 47231
Cabinet Back Cover	4835 432 97396
Cabinet Front	0014 688 90014
Crystal Bezel	4835 450 67136
Grille	4835 459 47048
Lens, IR	4835 381 17006
Nameplate	4835 459 17388

Figure 3-14. Television parts lists.

Placement Chart

The placement chart shows the top and bottom (when appropriate) of a circuit board, and shows the exact placement of each component. (*Figure 3-15.*) Also, placement charts also provide a grid map that has the component callout and the grid location to help you locate a component.

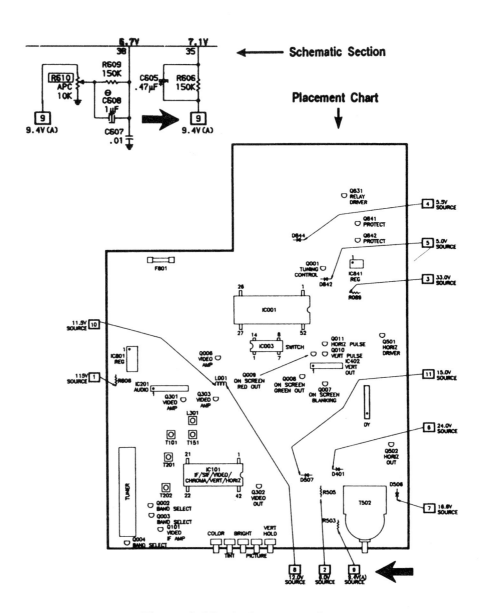

Figure 3-15. A placement chart.

Schematic

Figure 3-16 shows a portion of the type of schematic that you will use when troubleshooting and repairing televisions. The schematic is broken into stages, showing the components in each stage. If you need to jump to another schematic, the page and stage number of the new location are also included.

CHAPTER 3: THE BASICS OF TROUBLESHOOTING

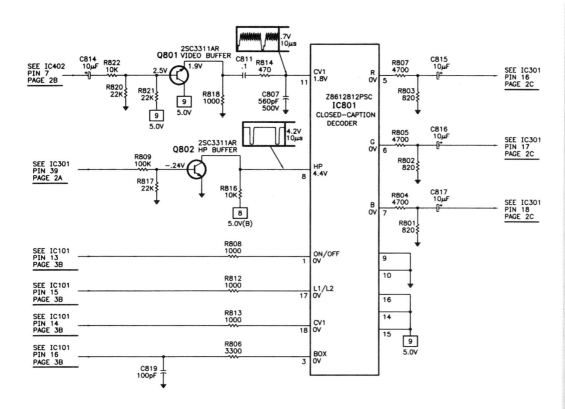

Figure 3-16. A portion of a television schematic.

Quiz

1. In the audio circuits, which troubleshooting method is best: signal injection or signal tracing?
2. What does SMT stand for?
3. Why use sharp probe points?
4. Why do you NOT use a soldering gun when working with MOS and FET devices?
5. What is a cold solder joint?

Key

1. Signal tracing.
2. Surface mount technology.
3. To reduce the chances of slipping off of the leads of components, and thereby prevent shorting or arcing.
4. The magnetic field generated by soldering can damage some MOS and FET devices.
5. An open connection on a PC board.

Chapter 4
TROUBLESHOOTING POWER SUPPLIES

Note: When you check voltages in the troubleshooting procedures, refer to the television's schematic for the correct voltages.

Chapter 4
Troubleshooting Power Supplies

A power supply is the circuit that supplies the energy to operate a television. It can be a separate assembly, or it can be integrated into the electrical system of the television. In today's televisions, the power supply components are usually complex. This chapter discusses the various power supplies in televisions and outlines some troubleshooting techniques you can use to locate a problem with the television's power supply. A block diagram of basic power supply stages is shown in *Figure 4-1*.

Also, some chassis require you to use an isolation transformer, shown in *Figure 4-2*, when you troubleshoot power supply problems.

Standby Power Supplies

Standby power supplies are located on the hot ground side of the power supply and secondary connections are usually cold ground. This power supply offers a means of turning the television on and off and using a remote control. These circuits are always on.

If you suspect that a standby power supply is faulty, use a voltmeter to test each component in the circuit. Then, place a jumper across the mode switch's emitter and collector terminals to see if the chassis turns on. Then, test each transistor and diode in circuit. If you find a leaky or open component, replace it.

Conventional Power Supplies

Most television circuits require a power supply that provides DC voltage. The basic unregulated DC power supply functions are transformation, rectification, and filtering. These functions are illustrated in *Figure 4-3*.

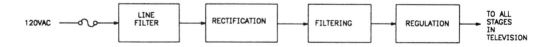

Figure 4-1. Basic power supply stages.

CHAPTER 4: TROUBLESHOOTING POWER SUPPLIES

Figure 4-2. An isolation transformer.

The transformation function input from the utility line is the 110V AC. Its AC output voltage, which can be lower or higher than the input voltage, is the input to the rectification function. The rectification function output is a DC voltage, but because it has large amplitude variations, it is called a pulsating DC. The filtering function reduces the large amplitude variations of the output to a DC voltage with only a small "ripple" voltage riding on it. This basic power supply is called an unregulated DC power supply because its output varies with the changes in the AC input voltage, as well as changes in the load on the power supply output. To help you understand the operation of this power supply, each of the functions is discussed below.

Transformer

A transformer is the component that performs the following two functions:
  1. Transforms the AC line voltage to the voltage value required to produce the proper AC voltage input to the rectification functions.
  2. Electrically isolates the electronic equipment from the utility power lines.

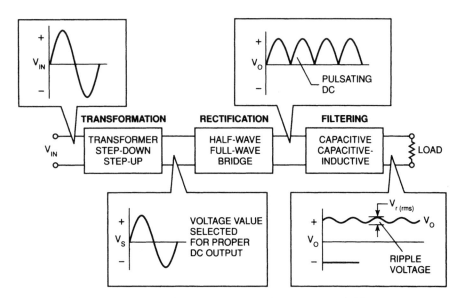

Figure 4-3. Transformation, rectification, and filtering of an unregulated DC power supply.
(Reprinted with permission from *Power Supplies: Projects for the Hobbyist and Technician*
© 1991, 1992, Master Publishing, Inc.)

As shown in *Figure 4-4*, the transformer consists of at least two coils of wire wound on the same iron core. The coil of wire receiving the input voltage is called the primary. The coil of wire receiving the output voltage is called the secondary. Many times there are two or more places in the secondary from which the outputs are taken.

The basic operating principle of a transformer is induction—the capability of a coil to store energy in a magnetic field surrounding the coil. As the current increases, the magnetic field increases. As the current decreases, the magnetic field decreases. The expanding and collapsing magnetic field induces a counter voltage in the coil that opposes the varying current that produces

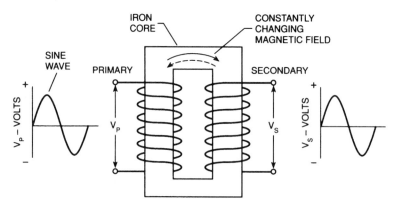

Figure 4-4. Transformer construction.
(Reprinted with permission from *Power Supplies: Projects for the Hobbyist and Technician*
© 1991, 1992, Master Publishing, Inc.)

CHAPTER 4: TROUBLESHOOTING POWER SUPPLIES

the original field. This opposition to AC is called inductive reactance, and its unit of measurement is the Henry.

Varying current from an AC voltage when it is applied to the primary creates a changing magnetic field in the iron core. This magnetic field, coupled to the secondary through the core, cuts into the secondary windings and induces an AC voltage in each turn of the secondary. Thus, energy is transferred from the primary to the secondary by varying the magnetic field without any electrical connection between them.

Because the energy transfer is accomplished only by the magnetic coupling between the primary and the secondary, the secondary and any circuits connected to it are isolated from the primary and any other circuits connected to the primary. This is very important for safety because the primary is connected to the high current supply of the utility line. Without such isolation, there is a serious shock hazard. Another advantage is that no DC connection exists between the circuit ground in the primary circuit, and the circuit ground in the secondary circuit.

In an unregulated power supply, the transformation function must provide an AC output voltage value required to produce the proper DC output voltage. This is easily accomplished in a transformer by varying the ratio of the number of secondary turns to the number of primary turns. The secondary voltage can be made less than or greater than the primary voltage simply by varying the turns ratio.

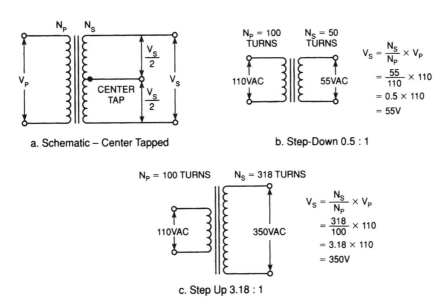

Figure 4-5. Diagrams of various types of transformers.
(Reprinted with permission from *Power Supplies: Projects for the Hobbyist and Technician* © 1991, 1992, Master Publishing, Inc.)

Knowing that the secondary voltage is the primary voltage times the turns ratio, it is easy to see how the transformer can be used to vary the voltage level of the AC voltage used in the unregulated power supply system.

Figure 4-5 shows several examples using schematic drawings for several types of transformers:
1. In the step-down transformer there are fewer secondary turns than primary turns, and the secondary voltage is less than the primary voltage. However, the current is stepped up.
2. In the step-up transformer there are more secondary turns than primary turns, and the secondary voltage is greater than the primary voltage. However, the current is stepped down.

Rectifier

Rectification converts an AC voltage into a DC voltage. A diode performs the rectification.

The rectifier acts as a one-way valve for electricity, as shown in *Figure 4-6*, because the diode lets electrons flow freely in only one direction. Forward-biased means that the anode voltage is more positive than the cathode voltage. Reverse-biased direction is the opposite of forward-biased, and the anode voltage is more negative than the cathode voltage. In reverse-bias, the electrons cannot flow as easily as in forward-bias.

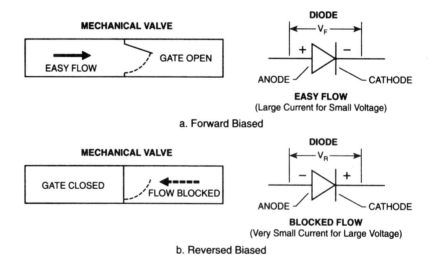

Figure 4-6. Forward-biased and reverse-biased rectifiers.
(Reprinted with permission from *Power Supplies: Projects for the Hobbyist and Technician*
© 1991, 1992, Master Publishing, Inc.)

CHAPTER 4: TROUBLESHOOTING POWER SUPPLIES

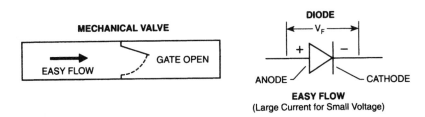

Figure 4-7. A forward-biased diode.
(Reprinted with permission from *Power Supplies: Projects for the Hobbyist and Technician*
© 1991, 1992, Master Publishing, Inc.)

Forward-Biased Diode (VF)

A forward-biased diode is like an electronic valve that's open: it allows the current to flow in only one direction. The resistance of a forward-biased diode is enough to develop a voltage drop across it. The most common diode is made of silicon and has a typical voltage drop of .5V to .7V. A germanium diode has a typical voltage drop of .05V to .2V. The silicon diode in *Figure 4-7* requires the anode to be .7V more positive than the cathode, to be forward-biased, and to allow current flow. If the forward-biased current exceeds the rated value in amps, it could result in permanent damage to the diode.

Reverse-Biased Diode (VR)

The maximum reverse-biased voltage that can be applied to a diode, shown in *Figure 4-8*, is called the peak inverse voltage (PIV). If this voltage is exceeded, the anode-cathode junction may break down and let a large current flow in the reverse direction, plus the diode will probably be permanently damaged.

Half-Wave Rectifier

The half-wave rectifier circuit is really one diode with an output of 60 Hz, and is used for low-current applications. A simple half-wave rectifier circuit has a

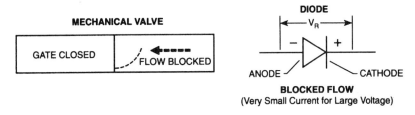

Figure 4-8. A reverse-biased diode.
(Reprinted with permission from *Power Supplies: Projects for the Hobbyist and Technician*
© 1991, 1992, Master Publishing, Inc.)

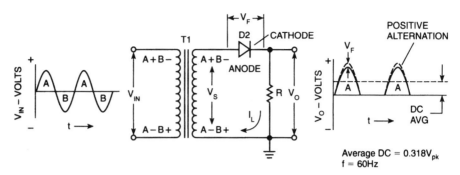

Figure 4-9. A half-wave rectifier circuit.
(Reprinted with permission from *Power Supplies: Projects for the Hobbyist and Technician*
© 1991, 1992, Master Publishing, Inc.)

diode connected in series with a transformer secondary output, as shown in *Figure 4-9*. The input primary voltage is a power line, 60 Hz, sine-wave voltage. The positive cycle alternation is labeled A. The negative cycle alternation is labeled B. The polarities of the primary and secondary voltages are noted for each alternation.

On alternation A, diode (D1) conducts because its anode is more positive than its cathode. A voltage equal to the secondary voltage minus the diode's forward-bias voltage (V_F) is developed across the load (R). Alternation B appears as reverse voltage (V_R) across the diode. To withstand this voltage, D1's PIV must be greater than the alternation B's voltage peak.

The output voltage is a series of 60 Hz, half-cycle alternations. The voltage is always in one direction and is known as pulsing DC. If the area under the positive pulses is averaged over a complete cycle, the average DC voltage is 0.318 times the voltage peak.

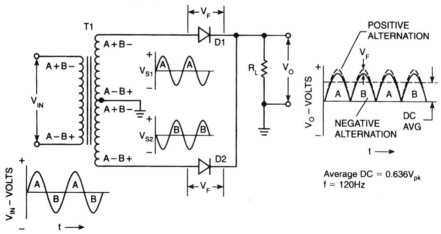

Figure 4-10. A full-wave rectifier circuit.
(Reprinted with permission from *Power Supplies: Projects for the Hobbyist and Technician*
© 1991, 1992, Master Publishing, Inc.)

Full-Wave Rectifier

The rectifier circuit shown in *Figure 4-10* is really two diodes and converts both alternations of the secondary voltage to a DC voltage. Therefore, it is known as a full-wave rectifier. A diode is in series with each secondary output and the center tap is grounded. The voltage between the center tap and each secondary output is equal in value, but 180° out of phase. When V_{S1} is positive, V_{S2} is negative.

D1 conducts during the A alternation, and D2 conducts during the B alternation. The center tap is a common return for the current from each diode. Because both diodes conduct current to the load in the same direction, the pulsating DC has positive half-cycles on both alternations of the secondary voltage. The output is pulsating DC at 120 Hz with an average DC voltage of 0.636 V_{PK}.

Only half of the secondary is used at one time. Therefore, the transformer's secondary output voltage must be twice that needed to provide the proper DC voltage. Also, each diode's PIV must be at least the secondary's (V_S) peak-to-peak voltage.

Bridge Rectifier

The bridge rectifier, shown in *Figure 4-11*, uses four diodes in a bridge network. One output terminal of the bridge network is a common ground for the return of the load current. The other output terminal is connected to the load.

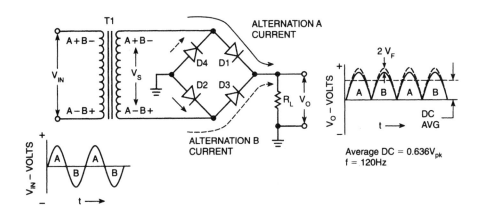

Figure 4-11. *A bridge rectifier circuit.*
(Reprinted with permission from *Power Supplies: Projects for the Hobbyist and Technician*
© 1991, 1992, Master Publishing, Inc.)

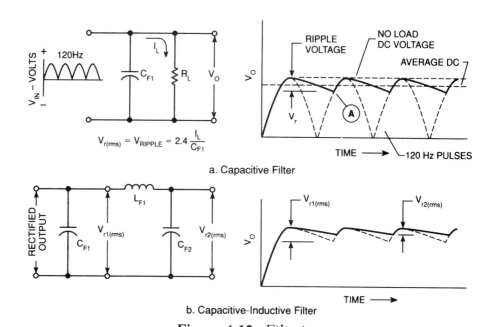

Figure 4-12. Filtering
(Reprinted with permission from *Power Supplies: Projects for the Hobbyist and Technician*
© 1991, 1992, Master Publishing, Inc.)

D1 and D2 conduct on the alternation A. D3 and D4 conduct on the alternation B. Both conducting paths deliver current to the load in the same direction. The pulsating DC output is the same as for a full-wave rectifier. The output DC voltage is the secondary voltage minus two forward diode drops. All of the diode's PIV must be greater than the secondary peak-to-peak voltage (V_s).

Filtering

The pulsating DC output after transformation and rectification is not a satisfactory power source for most electronic circuits. The filtering function, shown in *Figure 4-12*, smooths the output so that a nearly constant DC is available for the load. If there is excess ripple, the filter capacitor is not able to store enough charge to supply the required load and keep the output at a stable level. The ripple can cause the television to display symptoms such as hum or an unstable sync.

The pulsating DC output from the rectifier contains an average DC value and an AC portion called a ripple voltage. A filter circuit reduces the ripple voltage to an acceptable value. Resistors, inductors and capacitors are used to build filters. None of these components have amplification.
  1. Resistors oppose current and normally function the same way in DC
     or AC circuits by not varying with frequency.
  2. Inductors oppose current changes, and their inductive reactance

increases with frequency.
3. Capacitors oppose voltage changes, and their capacitive reactance decreases with frequency.

Capacitor

A capacitor is made of two conductive plates separated by an insulator called the dielectric, shown in *Figure 4-13*. The symbol found on a schematic is also shown in *Figure 4-13*. When DC voltage is applied across the plates, electrons collect on one plate and positive ions collect on the other plate. The difference in the electrical charges on the plates equals the voltage applied to the dielectric's lead. If the voltage is removed, the electrical charges remain in place and maintain the voltage difference between the plates. In this manner, the charge is stored by the capacitor.

This charge storage characteristic gives a capacitor in a circuit the effect of opposing voltage charges. For this reason, you can have two capacitors working against each other to smooth the ripple from the unregulated DC voltage. This is very important to the filtering function in DC power supplies.

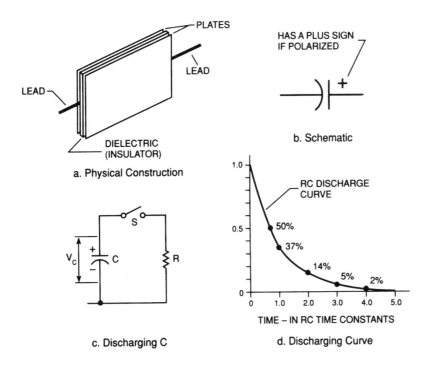

Figure 4-13. Capacitors.
(Reprinted with permission from *Power Supplies: Projects for the Hobbyist and Technician* © 1991, 1992, Master Publishing, Inc.)

The electrical unit of capacitance is a farad. A farad is a very large quantity so actual capacitors are usually rated in microfarads. One microfarad is 0.000001 (1×10^{-6}) farad.

Troubleshooting Conventional Power Supplies

To troubleshoot problems in power supplies, you can use the following equipment:
1. DMM with a rectifier tester or a VOM.
2. Capacitance tester.
3. Isolation transformer.

First, look at the most obvious areas that may cause problems:
1. A hot chassis.
2. AC or DC line input.
3. Power line cord and plug.
4. Power switches are turned off.
5. Blown fuse.
6. Series resistive fuse.
7. Thermal overload circuit.

Then, look at other areas that you think might fail.

A problem with the power supply can cause the following symptoms. The television:
1. Is totally dead.
2. Makes a humming noise.
3. Has a small picture.
4. Blows fuses when turned on.
5. Has a vertically and horizontally unstable picture (also called a "breathing" picture).
6. Has black bars that drift from bottom to top.

Defects in the power supply that cause these problems are shorted diodes or filter capacitors, open diodes or filter capacitors, dropping resistors, fuses or switches, leaky filter capacitors or a shorted transformer or choke.

For example, 60 Hz or 120 Hz hum can point to a faulty filter capacitor. Faulty filter capacitors also can cause vertical rolling and horizontal tearing. You can use an oscilloscope to locate a bad filter capacitor by looking at the waveform. You also can jump another capacitor across the one you suspect as being faulty to see if normal operation resumes. If so, then replace the

CHAPTER 4: TROUBLESHOOTING POWER SUPPLIES

faulty capacitor. Each time you jump across a suspected capacitor, disconnect the power cord and discharge the jumped capacitor in order to keep from damaging solid-state devices.

After you have checked the possible causes and they are not the source of the problem, look at the circuitry more closely. Before you check the television's internal circuitry, disconnect the television from the wall outlet. Just turning off the television is not enough. Then, follow the troubleshooting methods described in the section *General Techniques for Servicing Televisions*, in Chapter 3, **The Basics of Troubleshooting** to find the problem.

Locating Power Supply Problems

If a power supply fails, it might be due to a problem internal to the power supply or a problem induced by the operating circuit external to the supply. To check or test the power supply, perform these tasks:
1. Disconnect the circuit from the rest of the chassis. If the problem was
   not the power supply or damage to the supply components, the power
   supply output voltage should return to normal or somewhat higher.
2. Place an external load on the output to make sure the power supply functions properly under a full load.

If the power supply functions normally under a full load, the original problem is in another circuit.

If the power supply does not function normally, the problem is with the power supply.

If the power supply is defective, check the capacitors, resistors, diodes, transistors and inductors in the power supply. Use a DMM with a diode function to check the diodes and transistors. Also, remember that ICs are difficult to check. If the source of the problem is an IC, take careful voltage readings of all inputs to and outputs from the circuits in the IC, and of all components that connect the circuits. If the components that connect the circuits on the IC appear to be working correctly, the IC itself is probably faulty. The usual procedure in this case is to replace the IC.

The half-wave rectifier's circuits are normally protected with one or more fuses which are located on the main chassis. Fuses and silicon diodes present the most common problems in conventional power supplies. The fuse and diode can open if the components in the power supply in the connecting circuits are overloaded. When you replace a blown fuse, be sure to always use an exact replacement.

If the power supply contains a regulator with protective elements, there is probably no component damage due to the current overload because most three-terminal IC regulators are well protected internally. If there is no power supply voltage due to overload, the supply output should immediately return to normal when the external problem is repaired. If the regulator incorporates a thermal overload control that is tripped because of excessive power dissipation in the supply, let the power supply cool down before returning to normal operation after the source of the problem is repaired.

Other possible power supply failures include:
1. Loss of filtering due to capacitors that have changed.
2. Loss of output voltage because of a shorted output capacitor, open-circuited resistor or an open choke.

A failure in a power transformer, such as a shorted or open winding, would produce reduced output DC voltage or no voltage.

The main reason for IC regulator failures is too much dissipation in the internal series-pass transistors. If this happens, there will be no output voltage, but the input voltage will be normal. Also, if the output is shorted, there will be no voltage output because the regulator may have been shut down by the short-circuit protection provided internally.

Regulated Power Supplies

In most cases, a power supply must control its output voltage (or voltages) closely as its input voltage and output load are changed. This is called a regulated power supply. If the output was not regulated, the variations in its output voltage could become mistaken as signals within the system circuits being powered and cause errors, distortions, extra signals, and so on.

Most ICs require a precise power supply that controls the voltage levels within narrow limits. The supply must respond very quickly to peaks and dips in the current demand, because the ICs are adversely affected if certain voltage variations occur.

To design a regulated power supply, an unregulated supply, like those described in the section *Conventional Power Supplies* earlier in this chapter, is used and a regulator circuit is added to its output. The regulator can monitor

CHAPTER 4: TROUBLESHOOTING POWER SUPPLIES

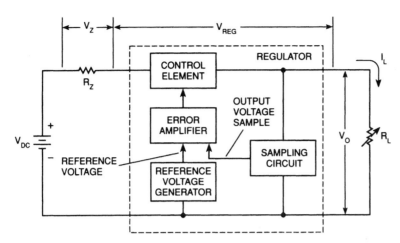

Figure 4-14. A regulated circuit.
(Reprinted with permission from *Power Supplies: Projects for the Hobbyist and Technician*
© 1991, 1992, Master Publishing, Inc.)

the power supply output and automatically make adjustments so the output voltage stays within defined limits. A simplified regulated circuit is shown in *Figure 4-14*.

To perform the required regulation, the regulator circuit varies the voltage (V_{REG}) to keep the output (V_O) constant as the load (R_L) changes. If the load (R_L) decreases, the load current increases. This tends to reduce the voltage output (V_O). However, the regulator reduces the voltage drop (V_{REG}) to offset the increase in current so that the voltage output (V_O) remains constant. Conversely, if the load (R_L) increases, which tends to decrease the load current, the regulator increases the voltage drop (V_{REG}) to keep the voltage output (V_O) constant. Similarly, the regulator increases or decreases the voltage if the rectified voltage drop (V_{REG}) increases or decreases.

The sampling circuit monitors the voltage output (V_O) and feeds a sample voltage to the error amplifier. The reference voltage generator maintains a constant reference voltage for the error amplifier regardless of input voltage variations. The error amplifier compares the sample output voltage to the reference voltage, and generates an error voltage if there is any difference between them. The error voltage is fed to the control element to control the value of the voltage drop (V_{REG}).

The control element essentially acts like a variable resistor which is in series with the rectified voltage, R_Z, and the load (R_L). When the rectified voltage or load (R_L) changes, the input to the control element from the error amplifier adjusts this variable resistance to change the voltage drop (V_{REG}) to hold the voltage output (V_O) constant.

Severe transient voltage increases on the unregulated rectified voltage can be a problem with IC regulators. If the problem is severe enough, the regulator can exceed the maximum voltage difference allowed between the regulator input and the output. In some cases, regulator failure puts the full unregulated voltage on the load circuit. ICs are very susceptible to failure due to excessive voltage and can fail in large numbers in complex circuitry as a result. In some cases, you will find a zener diode across the regulator output that acts as a clamp at a predetermined level if the supply voltage becomes excessive. Thus, the power supply shuts down without damaging the system.

Switched-mode power supplies can be a significant source of electromagnetic interference (EMI). Shielding and filtering are used to keep any generated EMI from escaping.

Zener Diode

Zener diodes, shown in *Figure 4-15*, are designed to maintain a fixed voltage across the diode junction and operate in the reverse breakdown region. When zener breakdown occurs, the strong potential across the depletion region causes a direct breakdown of the electron bonds, thereby freeing large numbers of electrons so a current can flow. The potential difference between the anode and the cathode is held constant and depends on how the diode was made.

Zener diodes are connected in circuits with their cathode as the positive terminal because they operate in the reverse-biased mode. Zener diodes are packaged in the same way as conventional semiconductor diodes. Small

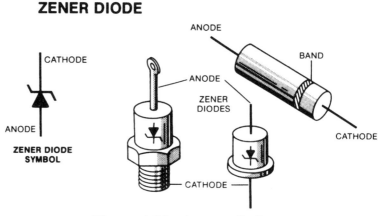

Figure 4-15. *A zener diode.*
(Reprinted with permission from *Power Supplies: Projects for the Hobbyist and Technician*
© 1991, 1992, Master Publishing, Inc.)

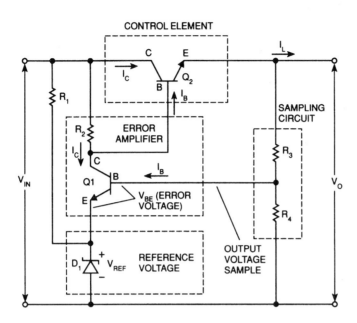

Figure 4-16. A feedback regulator.
(Reprinted with permission from *Power Supplies: Projects for the Hobbyist and Technician*
© 1991, 1992, Master Publishing, Inc.)

zener diodes are marked with a cathode band the same as conventional diodes. Large zener diodes are marked with the zener diode symbol, also shown in *Figure 4-15*. In addition to having a voltage rating (reverse breakdown), zener diodes have a power rating that must not be exceeded.

Zener diodes are widely used as voltage stabilizers in circuits because the voltage drop across the zener diode is relatively constant. These diodes behave like regular diodes when connected in the normal way. However, they are not usually used this way. When connected in reverse and sufficient voltage is applied with a suitable current-limiting resistor, the diode breaks down in the zener mode and holds the voltage constant across it.

One use for zener diodes is to supply the tuning varactors with constant voltage. If the zener diode is defective, then the tuning will become inoperative or unstable. Zener diodes can also be used as signal devices to limit peak-to-peak pulses.

When you test zener diodes, a faulty diode can cause a horizontal bar to float up on an oscilloscope screen. Also, a defective zener diode can cause erratic voltage changes. In this case, monitor the output voltage carefully. If the input is steady and the output changes, the zener diode is probably defective. A shorted zener diode will cause the power supply to have no voltage. An open zener diode will cause the output voltage to be high.

Series-Pass Feedback Voltage Regulator

The feedback regulator, shown in *Figure 4-16*, is a closed loop that feeds back a portion of the output voltage and compares it to a reference voltage. The difference between the two voltages determines the action that must be taken.

In the figure, the control element is an NPN transistor, Q2, which is connected in series between the input voltage (V_{IN}) and the output voltage (V_O). The load current (I_L) is the same as the collector current (I_C) to Q2. Therefore, all of the load current must pass through Q2. Collector current (I_C) cannot flow unless there is a base current (I_B) into Q2. This input to the base (I_B) controls the collector current (I_C) and the load current (I_L).

In this circuit, a decrease in the load current (I_L) tends to decrease the output voltage (V_O). A decrease in the output voltage (V_O) increases the base current (I_B) of the error amplifier which increases the collector current of the error amplifier. The increased collector current of the error amplifier reduces the base current from the control element. The reduced base current of the control element reduces its collector current, which increases its collector-emitter voltage. The increased voltage drop across the collector-emitter reduces the output voltage.

An increase in load current causes the opposite actions. Similar regulator control loop action occurs to keep the output constant if the input increases or decreases.

Troubleshooting a Series-Pass Feedback Voltage Regulator

When troubleshooting a series-pass power supply, use a voltmeter and measure the voltage at the output of the bridge rectifier to make sure the rectifier is operating correctly. If the voltage is not correct, you may have to repair the rectifier circuit. Check the reference voltage for the proper voltage; this voltage should remain constant. Then, check the error amplifier for proper voltages. Check the voltage on the base of the control element. Typically, the control element will be the faulty component.

Switched-Mode Power Supplies (SMPS)

Switched-mode power supplies, such as chopper, series-pass (linear), and self-oscillating, use switching regulators that can have a conversion effi-

CHAPTER 4: TROUBLESHOOTING POWER SUPPLIES

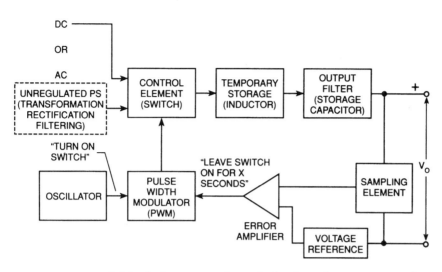

Figure 4-17. A block diagram of a switched-mode power supply.
(Reprinted with permission from *Power Supplies: Projects for the Hobbyist and Technician*
© 1991, 1992, Master Publishing, Inc.)

ciency of 85% or more. This efficiency results in lower power dissipation and smaller size components for a given power output. Other advantages of switched-mode power supplies are that they can:

1. Operate over a wide range of current and voltage.
2. Be used for the control element.
3. Have lower input voltage than output voltage.
4. Have output voltage that is opposite in polarity from the input voltage.
5. Have no off state, only on or standby.

Before beginning the troubleshooting procedures, make sure you are using an isolation transformer.

Switched-Mode Power Supply Operating Principles

Figure 4-17 shows a block diagram of a switched-mode power supply. There are some similarities between switching systems and the regulated power supply system discussed earlier in this chapter. The difference between a switched-mode and linear system is the action of an inductor used for temporary energy storage and how the control element is controlled to provide regulation.

If an AC source is used, the transformation, rectification and filtering circuits that provide a DC input voltage to the regulated power supply systems serve as a switching power source in switched-mode power supplies. If an unregulated DC source is used, an input filter may be required for ripple or noise reduction, or for stability.

In regulated power supply systems, regulation is accomplished by varying the voltage drop of the control element. In switching systems, this function is accomplished by rapidly turning the control element on and off, and by varying the ratio of on time to off time. Unlike the series-pass control element, there is no linear operating state. The control element is either completely on or off.

When the control element is on, energy is pumped into the inductor's temporary storage element in sudden bursts. When the control element is off, the stored energy is directed by a diode into the capacitor to supply the load as needed. The sampling element, reference-voltage, and error amplifier work in an identical manner to those in a linear supply. However, the output of the error amplifier is used directly to determine the length of time for the sudden bursts.

Compared to the linear system, the switched-mode system has three new circuits: the oscillator, the pulse-width modulator (PWM) and the temporary storage element inductor.

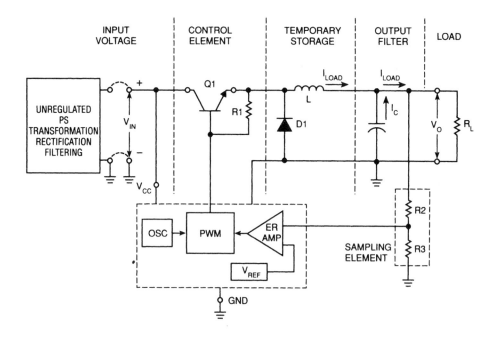

Figure 4-18. A step-down switching power supply.
(Reprinted with permission from *Power Supplies: Projects for the Hobbyist and Technician*
© 1991, 1992, Master Publishing, Inc.)

Regulators in Switched-Mode Power Supplies

There are three types of regulators:
1. A step-down regulator, also called a buck, is used when the required
   voltage is lower than the input voltage.
2. A step-up regulator, also called a boost, is used when the required voltage is higher than the input voltage.
3. A flyback regulator, also called an inverter regulator, is used when a
   regulated output voltage of opposite polarity is required.

Figure 4-18 shows a functional block diagram of a complete step-down switching power supply. Each of the individual functions is discussed below.

Input Voltage

The input voltage for a switched-mode power supply can vary over a surprisingly wide range while still maintaining good conversion efficiency. An AC source is indicated for V_{IN}. All the same design considerations apply for the transformation, rectification and filtering as for the unregulated supply described in the section *Conventional Power Supplies*, earlier in this chapter. The ripple voltage filtering, which takes care of the 60 Hz and 120 Hz noise, can be somewhat less because of the output filtering of the switching regulator.

Control Element Switch

The control element switch is a switching power semiconductor. It must have low V_{ON} across it and must have fast switching times. It can be a single NPN transistor or a power FET, or it can be a combination NPN and PNP for higher gain operation. Regardless of the type transistor used, it is turned completely on (lowest resistance) during the ON time. Then, it is turned completely off (highest resistance) during the OFF time. It is used as if it were a relay, but operates at a very fast rate—from 20 kHZ to 100 kHz. A 100 kHz wave has a period of 10 microseconds (0.00001 second).

Catch Diode

When the magnetic field of the inductor begins to collapse and release its stored energy, the energy must be contained and channeled in a useful direction. A diode called the catch diode performs this task in each of the circuits. It directs the stored energy into the output filter capacitor. The one-way conduction of the diode is used to provide the proper circuit connection when the induced voltage across the inductor is the proper polarity. Due to the high switching frequencies in these supplies, the diode must have a very low forward voltage and a very fast switching time and recovery time. A Schottky diode is an ideal diode for this application.

Inductor

The inductor must have the proper inductance and must not saturate during its operation. The core must have a volume to handle the power required. If it saturates, it loses its inductance and its ability to efficiently transfer energy to the output. Remember, inductance is determined by the core material and the number of times the wire is wound on the core. Ferrite and powdered iron cores are usually used for switching power supply inductors. When iron laminated cores are used, they have core loss at higher frequencies, which lowers supply efficiency.

Filters

As in the series-pass regulator, the output filter capacitor stores energy to be used by the load. Because the output ripple frequency is much higher in the switched-mode regulator, the output filter capacitor usually is a much smaller value than that for the series-pass regulator. The input filter must reduce the 60 Hz or 120 Hz ripple to acceptable values, and it must keep the switching-frequency ripple from the input to keep the system stable and noise free.

Troubleshooting Switched-Mode Power Supplies

Normally, the switched-mode power supply circuit will turn off if it develops a short. If the circuit is open, the power supply will continue to output voltage but not current. In some televisions, if the power supply samples the high voltage and finds that the voltage exceeds its limits, or if the sensing circuit is faulty, the power supply will turn off. Also, some power supplies have an undervoltage lock circuit that will turn off the power supply if the AC or regulated outputs are too low.

To troubleshoot a SMPS, follow these steps; make sure you are using an isolation transformer:
1. Inspect the circuit carefully for defects and/or problems, such as burned components.
2. Make sure the standby power supply is operational, if necessary.
3. Disconnect the horizontal output transistor. This will allow service

CHAPTER 4: TROUBLESHOOTING POWER SUPPLIES

to the switched-mode power supplies with less loading. Then, start disconnecting the other loads one at a time, and check for shorted loads after they are disconnected from the transformer.

4. Remove the signal to the base of the switching transistor by disconnecting the transistor. This prevents the power supply from outputting power as you check the other circuits.
5. Monitor the DC voltage at the collector of the switching transistor as you slowly increase the AC input voltage. If the DC does not increase, notice the AC current draw. If there is no current, and no or low DC voltage, the unregulated power supply is open, there is an open safety resistor, or there is an open primary in the switching transformer. If there is no or low DC and the ammeter current increases, there is usually a shorted switching transistor.
6. Make sure the base has enough voltage from the standby power supply to start the SMPS.
7. Check the feedback circuit by turning the circuit on.
8. Connect a variable current limited power supply.
9. Connect a voltmeter to the input of the base circuit. If the circuit is operating properly, when you increase the voltage, you will see a corresponding change at the input of the base.
10. Repeat steps 6 through 9 for all of the SMPS circuits and different transistors until you locate the defective circuit or component.
11. If you have not found the defective circuit or component, you can reconnect the base voltage and continue troubleshooting using your normal voltage analysis methods.

Scan-Derived Power Supplies

Scan-derived power supplies use the horizontal output voltage being wasted as a supply potential. This energy is produced by the horizontal output because of the frequency at which it is pulsed and because it is oscillatory energy. This energy source can almost power the entire chassis, including the CRT.

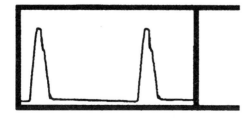

Figure 4-19. A scan-derived signal waveform.

Note: *Use care when measuring the voltage at the horizontal output collector unless you first disconnect the collector from the circuit. Many DMMs and oscilloscopes can be damaged by the high peak-to-peak horizontal waveform on the collector.*

Note: *When you use a variable line transformer, start with a low voltage and slowly increase the voltage.*

The energy from the horizontal output eliminates the necessity for power transformers, and elaborate low-voltage supply circuits. This reduces the amount of heat generated by a power transformer, making televisions cooler to operate, lighter chassis, and reduces the cost of television production.

Figure 4-19 shows the scan-derived signal waveform or pulse. The pulses are produced at retrace time, and the negative base line between the pulses is produced during scan time. If a rectifier is inserted in the circuit to allow conduction of positive pulses, the result is pulse rectification. If the rectifier conducts on the negative portion of the wave, the result is scan rectification.

The pulse voltage is usually greater than the DC voltage. Therefore, replace any diodes in the circuit with exact replacements. In other words, the diode cannot be reversed because the secondary windings would load down and the scan-derived circuit would not operate properly. Also, replace any rectifiers in the circuit with exact replacements that are capable of high-speed switching.

Troubleshooting Scan-Derived Power Supplies

If the television turns on briefly, then turns off completely, you may hear a short burst of audio when the television turns on. When this happens, the voltage regulator circuit senses the overload and turns off the circuit, which turns off the television.

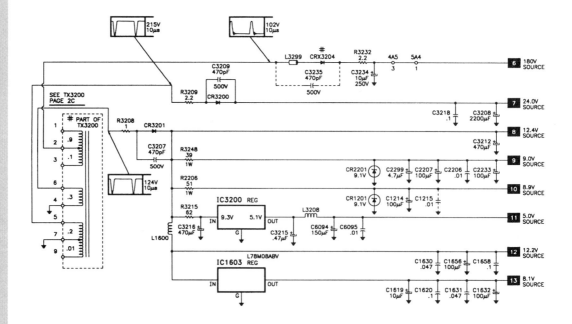

Figure 4-20. Scan-derived circuits.

CHAPTER 4: TROUBLESHOOTING POWER SUPPLIES

In newer televisions, the startup and secondary power supplies may be supplied by the scan-derived circuits, as shown in *Figure 4-20*. So, if the horizontal output circuit is not working properly, the chassis will not power up.

If this happens, follow these steps to locate the source of the problem:
1. Check all fuses.
2. Check the chassis for burn marks, including connections and components.
3. Check the boost source. If there is no boost, the source of the problem is in the power supply circuit. If there is a horizontal output signal, use signal tracing until you locate the faulty port.

Use a variable line transformer to reduce the AC input to 75V. Then, slowly increase the voltage until you detect the problem.

If one of the sources for the scan-derived circuits is not working correctly, measure the voltage at all connections and look for excessively low values. Do not turn off any DC voltage to the horizontal circuits. A low voltage measurement can indicate a defective or overloaded circuit. Shut down the chassis and disable the suspected power source. If the voltage on the other circuits increases, the scan-derived circuit you removed is defective. Troubleshoot all circuits supplied by this voltage.

Normally, if the flyback is faulty, all sources are bad or you will see arcing. Check for leaky or shorted diodes. Also, look for burned areas and poor solder joints.

Quiz

1. What are the basic unregulated DC power supply functions?
2. What is the difference between a forward-biased diode and a reverse-biased diode?
3. What diode is designed to operate in a reverse-biased condition?
4. What type of power supply supplies power to the circuits before the set is turned on?
5. What are power supplies removed from the flyback called?

Key

1. Transformation, rectification, filtering.
2. Forward-biased allows current to flow through; reverse-biased does not.
3. Zener diode.
4. Standby power supply.
5. Scan-derived.

Chapter 5
TROUBLESHOOTING VIDEO CIRCUITS

Note: The circuits shown in this chapter are commonly-used circuits. There are circuit variations used by manufacturers that perform the same functions, but may have different stages and configurations than the ones shown.

Chapter 5
Troubleshooting Video Circuits

Figure 5-1 shows a block diagram depicting the basic video stages. The shaded areas on this block diagram are the areas in the television that are involved with producing video:

1. IF amplifier.
2. Video detector.
3. Video amplifier.
4. Audio/video switching.
5. Comb filter.
6. Delay line.
7. Video processing.
8. Last video amplifier.
9. Chroma processing.
10. PIP.
11. How a CRT works.

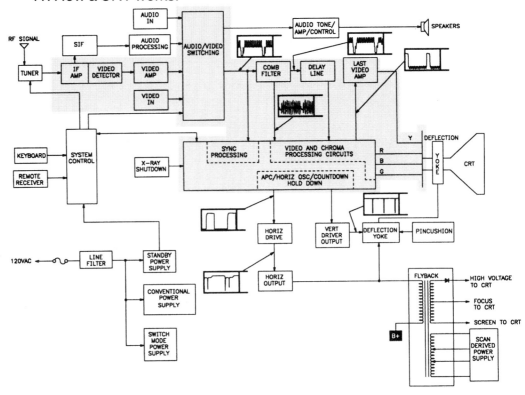

Note: When you check voltages in the troubleshooting procedures, refer to the television's schematic for the correct voltages.

Figure 5-1. A block diagram showing the basic video stages (shaded).

CHAPTER 5: TROUBLESHOOTING VIDEO CIRCUITS

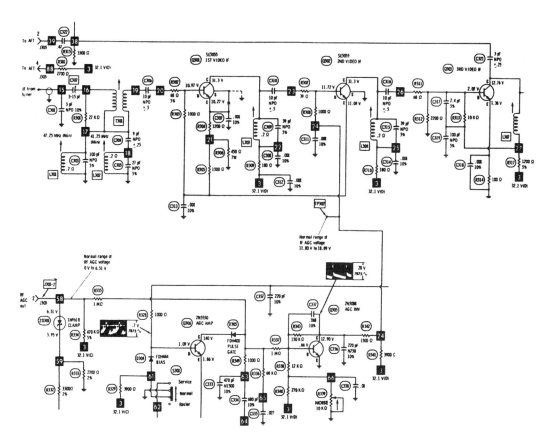

Figure 5-2. An IF amplifier connected to an AGC.

If, when troubleshooting you are measuring voltages equal to or more than 1000V, use a high voltage probe. Very high RF voltages may burn out an oscilloscope and other equipment designed for lower voltages.

Troubleshooting the Video IF Amplifier

The purpose of the IF amplifier, shown in *Figure 5-2*, is to provide the required increase to the strength of the signal received from the tuner. The AGC is coupled to the IF amplifier and monitors the average strength of the sync signal pulses. The IF amplifier is usually three transistors linked together or is internal in an IC.

Since the composite video, audio and sync signals pass through the IF amplifier, a failure in this stage would affect all three signals. When you test the IF amplifier, measure all voltages on the IC using a voltmeter. *Figure 5-3* shows where in the circuit to test for the signals.

Note: Troubleshooting the IF amplifier and the video detector is very difficult at times. Measuring the inputs, outputs and associated components is the only test that can be made. If any measurements are in question, substituting the IC may be the only recourse.

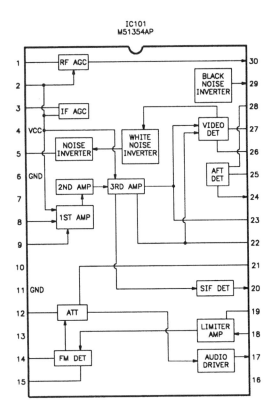

Figure 5-3. A diagram showing IC function.

When troubleshooting problems in the IC, measure voltages into and out of the IC itself. Also, measure voltages into and out of any components that are connected to the IC.

Figure 5-4 shows where to measure voltage in the IF amplifier. If the problem is in the first stage of the IF, the sound and picture are affected. However, some of the signal usually will pass through. Therefore, the symptoms may be:
1. Very weak video, or snow in the picture.
2. Some scanning raster.
3. A weak hissing sound that can be increased or decreased using the volume control.
4. No program sound.

The symptoms of a defective second stage are similar to the symptoms of the first and third stages. When you test the second stage, change the channel. If you notice any change in the picture, the IC is probably working properly.

Because the video detector is closely linked to the third stage, it is often difficult to tell if there is a problem in the IF amplifier or the video detector. If the problem is in the third stage, there is no audio or video; however, there

CHAPTER 5: TROUBLESHOOTING VIDEO CIRCUITS

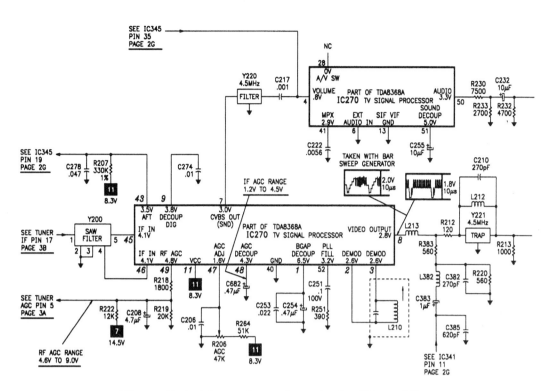

Figure 5-4. Measuring voltage in the IF amplifier. The pins in italics are the ones being examined.

is a scanning raster. This is very similar to the symptoms produced when there is a faulty video detector. There will be a soft background hum and maybe some background noise which you can control with the volume control.

Now, suppose you have a television in which the raster appears to be normal, there is some slight hissing when the volume is full, but there is no picture or sound. Changing channels does not produce a difference.

Check the signal at the last stage using a voltmeter. Measure the power supply voltage. If any of the voltages are suspect, substitute a new IC for the faulty one and normal operation should resume.

Troubleshooting the Video Detector

In newer televisions, the video detector, shown in *Figure 5-5*, is the last stage of the IF amplifier circuit. This circuit:
1. Converts the amplified signal from the IF amplifiers.
2. Separates (demodulates) the signal into a video signal with audio IF signals.
3. Sends the signals to the video amplifier.

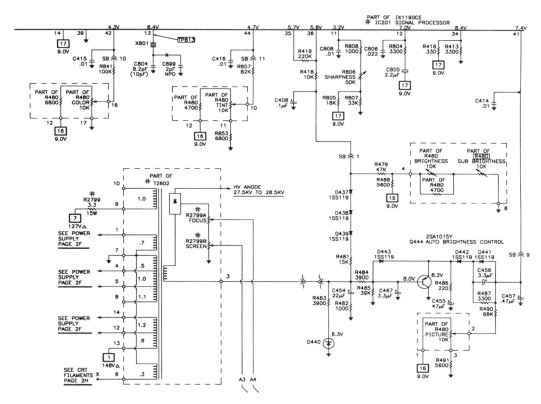

Figure 5-5. A simple schematic clip showing all of the controls, with the names called out.

Troubleshooting Symptoms

In older televisions, the video detector uses a diode to detect the input signal from the IF amplifier. (See *Figure 5-6*.) If you have to replace this diode, make sure you use an exact replacement and insert the replacement in the same polarity. If the polarity is reversed, the resulting picture will be a negative of the original (dark is light, and light is dark), as shown in *Figure 5-7*. In newer televisions, signal detection is accomplished by a synchronous detector that is part of an IC.

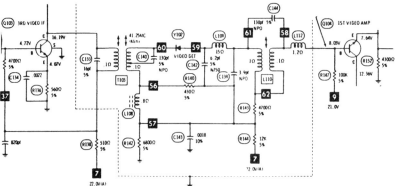

Figure 5-6. A diode used as a video detector.

Figure 5-7. A negative picture, indicating reversed polarity.

The symptoms of a faulty video detector are much like those caused by a faulty third transistor in the IF amplifier. There is no audio or video, but there is a scanning raster. Also, if the video detector is faulty, there will be a soft background hum and maybe some background noise that you can control with the volume control.

If the video detector's output signal decreases, the quality of the picture decreases. Use signal injection to test the video detector and measure the signal voltages with an oscilloscope. If you inject an IF output signal in the input of the video detector and a signal appears on the output using an oscilloscope, the video detector circuitry is probably good. If you inject a signal at the detector's output, and a signal appears on the CRT, the detector is faulty.

Troubleshooting the Video Amplifier

The video amplifier circuit increases the signal from the video detector. Then, the color, sound IF signal and the sync signal pulses are separated from the picture information. The sound IF signal is sent to the sound IF amplifier. The sync signal pulses are sent to the sync circuits.

If you use an oscilloscope, look for the composite video signal at the first transistor. (See *Figure 5-8.*) If you use signal injection, disable the video amplifier and inject the signal on the output of the video amplifier.

Note: In newer televisions, the sound IF amplifier is internal to the video IF IC, and is not accessible for testing. Only the output to the 4.5 MHz trap is accessible.

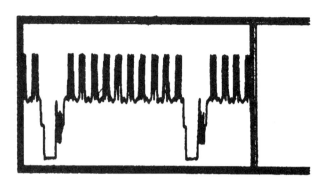

Figure 5-8. Waveform of a composite video picture at the first transistor.

In older televisions, the video amplifier circuits have three or four linked stages. (See *Figure 5-9*.) A failure in any one stage can prevent the signal from passing through correctly. However, even if one stage is completely dead, some of the original signal can still feed through. The symptom to look for is a very weak picture on all stations, not just one station. However, when receiving by antenna, some stations may appear to be stronger than others because of the varied strengths of the signals received by the antenna. Cable stations appear more uniform because of the uniform strength of the transmitted signals.

If the first stage in a three-stage video amplifier fails, the sound, video and synchronization can be affected. The audio, color and sync signals are most often taken off before the second stage. Therefore, failure after this point will disrupt the video, not the other three signals. For example, picture instability, such as rolling or tearing pictures, points to a failure in the first stage because the sync pulses are disrupted. If this first stage fails, the sync pulses can be degraded so that they cannot hold the picture vertically and horizontally. However, if the sound, color and raster are normal, the first stage is working correctly.

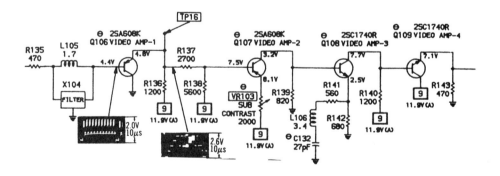

Figure 5-9. The linked stages of an older video amplifier circuit.

Suppose the symptoms you see are normal audio, no video, and faint raster? When you check the video amplifier circuits using a voltmeter, you may find that the second stage failed because of an emitter-to-collector short. Because of the short, the collector voltage at the second stage dropped. When the voltage dropped, the current dropped, and the stage shut off. This decrease in output voltage caused the picture brightness to decrease.

Another symptom of a faulty video amplifier are audio signals that show up in the video. The herringbone pattern of the signal seen on the screen can be caused if the 4.5 MHz audio trap circuit is misaligned or if there is a shorted capacitor in the circuit. If the capacitor is open, the video gain is reduced at high frequencies.

When you troubleshoot the video amplifier circuit, measure the input and output voltages using a voltmeter or measure the waveforms using an oscilloscope. Because the transistors are linked, the bias on each transistor is directly affected by the preceding transistors. Therefore, the collector bias voltage on the output of the first transistor is the same as the base voltage on the second transistor. It is important to check each transistor in the chain to determine which one caused the problem. Also, check them in order.

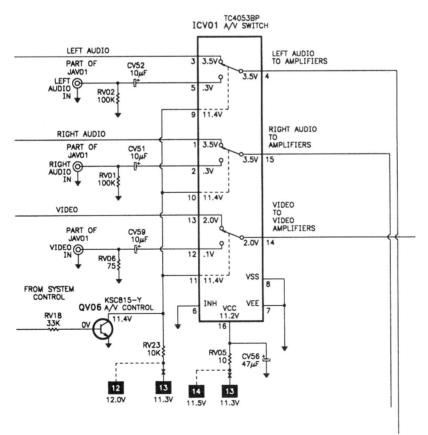

Figure 5-10. An A/V switching circuit.

The video amplifier in newer televisions is normally a single transistor amplifier or buffer that goes to the SIF circuit and an audio/video (A/V) switching circuit. After the A/V switching circuit, the signal goes to the video processing stage typically internal in an IC. The symptoms previously described can still apply. However, the only way to test the IC may be by signal injection. After the A/V switching circuit, use a voltmeter to measure all pins on the IC and a scope to measure all waveforms. If any voltages are in question, substituting the IC may be required.

Audio/Video (A/V) Switching

A/V switching circuits are used to switch to an alternate audio or video source. For example, a VCR video can be watched through the A/V inputs on the receiver, or a second video signal can be supplied to the PIP circuits of a PIP feature.

A switching circuit is normally an IC controlled by the microprocessor control circuit. Its purpose is to switch the proper audio and video signals to the processing circuits. Video switching problems are most often located in the system control or A/V switching circuits. (See *Figure 5-10*).

Comb Filter

The color signal is passed through a narrow bandpass filter that separates the chroma signal information from the luminance information. The comb filter is typically adjustable. To adjust the comb filter, connect an oscilloscope to the input of the last video amplifier. Adjust the potentiometer or adjustable coil for a minimum chroma component in the luminance signal.

Delay Line

The luminance portion of the video signal needs to be slowed down to allow for synchronization with the color signal. This is done through the delay line. A delay rarely fails, but if it goes bad it will produce either a faulty luminance waveform or no luminance waveform for the video processing circuits.

Picture Adjustment Controls

Note: To properly adjust the these controls, proper service literature should be obtained.

The following sections discuss the picture controls, which are located in the video processing circuits. These controls are in or around the video processing IC, and are rarely faulty. There are two main types of picture controls: The digital control and the analog control.

CHAPTER 5: TROUBLESHOOTING VIDEO CIRCUITS

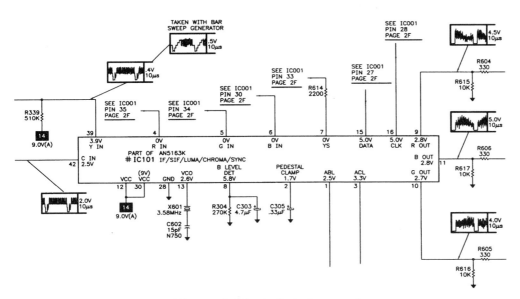

Figure 5-11. A digital control.

Figure 5-11 represents a digital control. The picture control settings are controlled internal to the IC101; however, the system control IC sends information to IC101 via the serial data line at pin 15. Use an oscilloscope to look for data transfer on pin 15—there is no other practical troubleshooting test to perform. Check for proper chroma and luminance waveforms at the inputs of IC101, then check all DC voltages on IC101 and IC001 (system control IC). Compare the voltages to those depicted in the PHOTOFACT schematic to determine which IC could be faulty.

Figure 5-12 represents an analog control. The picture settings are determined and controlled by the system control IC (IC345). When a picture control is adjusted, the voltage on the appropriate pin of IC270 (27, 29, 30, 31 and 32) will change to a new DC level. This is called stepped voltage control. The process that occurs is a pure analog action, determined internally in the system control IC (IC345). IC270 can be tested by using a variable DC supply and applying a bias to the appropriate pin of IC 270.

Since the mid-1980s, these two types of picture controls have become very common. The way the system control determine the settings is manufacturer dependent; however, there are chassis being made that still use the potentiometer to determine settings. This is also an analog-type control circuit. The following sections deal with these controls and how they affect the video signal.

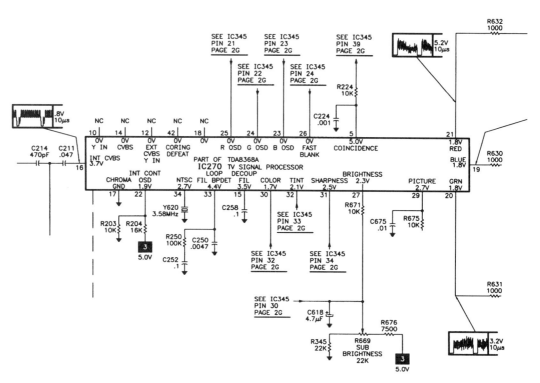

Figure 5-12. An analog control.

Peaking Coil

The video peaking coil is located in the video processing circuits and is resonant to high frequency signals. This means that high frequency signals that might be lost are boosted by the resonant circuit, resulting in amplified high frequency signals. The result is overshoot—light edges around dark images and dark edges around light images, producing a sharper image.

If the image is too sharp (harsh) or if the light to dark areas are washed out, like the pictures in *Figure 5-13*, use the television's sharpness control to adjust the frequency of the circuit. However, if you adjust the sharpness control and the problem persists, there could be a problem with alignment or with the CRT.

Sharpness Control

The sharpness control is connected to the video processing IC and varies the level of high frequency signals to the video amplifier. To accomplish this, the sharpness control causes some of the high frequency signals to go to ground, thus reducing the harshness of the picture. Also, the scanning raster (snow) and interference from other signals are in the high frequency range. If you adjust the television's sharpness control so that these high frequency signals are grounded, these signals do not interfere with the normal picture.

Figure 5-13. One picture (left) is too harsh and the other (right) is too dim.

If you see interference from other signals, and cannot reduce the interference using the sharpness control, check the video IF circuits.

Contrast and Picture Controls

The contrast control component varies the amount of white-to-black level of the video signal. (See *Figure 5-14*.) This control varies the signal gain on one of the video amplifiers by changing the resistance in the emitter circuit or the amount of signal voltage on the amplifier's base terminal. It does this with a voltage divider.

When the contrast control is not working correctly, the dark and light images are not distinct and appear to blend. Also, the picture control circuits on larger televisions usually control both contrast and color, and make the overall picture smoother and less grainy. When the picture control is not working properly, the images will look like those in *Figure 5-15*. If you adjust the contrast using the contrast and picture controls and the problems persist, check the video processing circuits, and components associated with sub-contrast and sub-picture level controls.

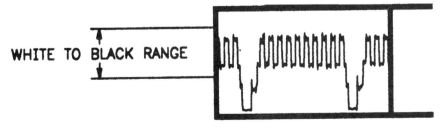

Figure 5-14. The contrast control varies the amount of black-and-white level of the video signal.

Figure 5-15. Contrast and graininess problems.

Brightness Control

The brightness control component adjusts the circuits that affect the DC bias, on the CRT, thus adjusting the amount of current to the electron guns. The picture can be very dark when the current is limited, or very bright when the current is increased.

Most televisions have an automatic brightness control circuit usually called a brightness limiter. If you use the brightness control and the picture does not improve, check the CRT board and circuits or the CRT itself.

Vertical and Horizontal Blanking

Vertical and horizontal blanking turn off the CRT during retrace so that retrace lines do not appear on the screen when the electron guns move from the end of a line to the beginning of the next line, and from the bottom right of the screen to the top left. The vertical and horizontal synchronizing pulses, which occur at the end of each line, cause a blank period to occur in the video signal between the lines. Failure can cause a very bright, washed-out picture with bright diagonal lines across the screen.

Last Video Amplifier

The last video amplifier is the last place to scope the video before it goes to the CRT board. Sometimes this is done in the video processing IC. However, if you scope a proper waveform at the input of the last video amplifier and the picture has no luminance, the last video amplifier needs to be checked. (See *Figure 5-16.*)

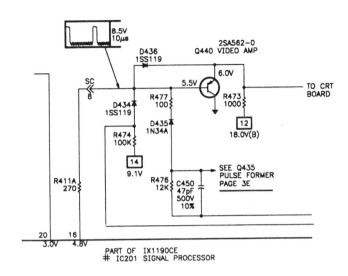

Figure 5-16. The last video amplifier.

Chroma Processing

When the original signal is produced in the television studio, the camera captures the color and sends the signal to a modulator. The modulator circuit combines the three color signals into one color signal. This circuit also produces a color sync signal. The color signal is then called a color modulated signal. The color sync signal (also called the color burst) is an unmodulated 3.58 MHz carrier.

A television's color processing circuits recreate a color picture and produce the color picture on the screen. Essentially any color can be created from the primary colors: red, blue and green.

When the signal is received, the color sync signal turns on the 3.58 MHz oscillator and synchronizes it with the signal being received from the station. The signal from the oscillator is output to the color killer which turns on the color IF amplifiers so that the signal can be amplified. After amplification, the signal is passed to the color demodulator circuits. In the demodulator circuits the original red, green and blue signals are separated and sent to the red, green and blue color amplifiers. The color amplifiers combine the three color signals with the black-and-white information (luminance), then pass the signals on to the three electron guns which produce the color picture on the screen.

When these circuits malfunction, the result can be no color, color that is too intense, loss of one color, one or all wrong colors, or loss of color synchronization. For example, if the demodulator circuits fail, you might see no color,

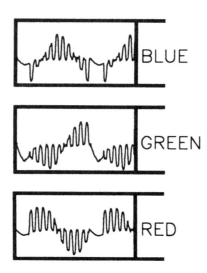

Figure 5-17. Red, green and blue waveforms.

an incorrect tint or intensity, or intermittent color, tint or intensity. *Figure 5-17* shows the red, blue and green color waveforms to expect when testing the color circuits.

Automatic Kine Bias (AKB)

Automatic kine bias (AKB) is circuitry used to adjust the beam current of each electron gun individually. This circuitry is intended to prolong the life of the electron guns and give optimum color setup. The setup is done only once in the lifetime of the CRT, or until a malfunction occurs within this circuit. The AKB IC samples the vertical blanking pulse for timing, and sources a small bias current to the cathode drivers to draw beam current independently on the electron guns. The AKB IC senses and adjusts with every vertical blanking pulse. For an optimum color screen setup, this is critical but relatively easy to set for most standard chassis.

No Color

If there is no color, the problem can be in the IF amplifier, the color killer, the oscillator, or the color amplifier. If any of these circuits are defective, color signals cannot be processed and there will be no color. When you test these circuits, test the color killer first.

Color Too Intense

If the color is too intense, try to readjust the picture using the color intensity controls. If this does not clear up the problem, the color amplifier may have too much gain. Use an oscilloscope to look at the input and the output of the color

CHAPTER 5: TROUBLESHOOTING VIDEO CIRCUITS

amplifier. Then, use a voltmeter to check the color killer and the automatic chroma control (ACC).

Losing One Color

[text partially obscured by receipt] ne color, one of the color guns in the CRT [...] g the proper signals. The resulting picture [...] lor. Check the color controls first. If adjust- [...] re the picture, the color video amplifier or [...] be faulty. For example, if the green color [...] appear in reds, blues and purples. When [...] check the CRT. Then, use an oscilloscope [...] k the color demodulator circuits and check [...] the missing color on the CAT circuit board.

One Color is Incorrect

[text partially obscured by receipt] just the picture using the tint control. If this [...] might be a problem with the color signal [...] shift of one color signal. Use an oscillo- [...] the output signal to the color signal de- [...] remember that it must have the correct [...] ese waveforms represent the pure color [...] oved.

All Colors are Incorrect

If all colors are incorrect, but there is color present, the tint control might be adjusted improperly. The tint control actually shifts the phase of the color oscillator output signal. If the tint control does not correct the problem, the color amplifier may be faulty or the sync is slightly off frequency with the burst. Use a voltmeter to test the color amplifier. Check also the CRT board and the three video output stage circuitry.

Loss of Color Sync

When the color sync is faulty, the color and the luminance are not synchronized, and the result is a chaotic picture. The picture may look like color rainbows within a black-and-white picture.

The horizontal oscillator frequency must be correct—15,735 Hz—for the luminance to be in sync. Likewise, the 3.58 MHz color oscillator frequency must be correct for the color picture to be in color sync. If the color sync is lost, check the

Figure 5-18. A picture with a PIP window.

signals from both the horizontal oscillator and the 3.58 MHz color oscillator. Use an oscilloscope to check the waveform from the color oscillator. Most chassis use a 3.58 MHz oscillator crystal, which can be defective.

Then, check all of the voltages passing through the color processing circuits and the components in the color processing circuits. For example, in the oscillator circuit, look carefully at the transistors. Not only does the oscillator have to work correctly, it also must be locked in phase with the color burst in the signal from the television station. If the color sync amplifier is not working properly, phase lock might be lost. If all of the chroma is processed inside an IC, replace the IC.

NTSC Color System

The NTSC—National Television System Committee—color system requires phase control of the transmitted color signals so that there is the correct relationship between the color burst and the color signals. If this phase is not controlled, color shifts appear in the signals, resulting in poor color picture quality. NTSC is a national standard that is enforced in the United States. Some countries have similar standards that are not compatible with NTSC.

The NTSC color bar generator is a good signal source when troubleshooting problems in televisions.

Picture-in-Picture (PIP)

Televisions with a PIP circuit, shown in *Figure 5-18*, let you view more than one picture at the same time. One picture will appear on the main screen, and another picture will appear in a small picture window on the screen. Some televisions let you show multiple small picture windows. Also, you can specify where on the screen the small window appears. You can swap the pictures so that the picture that appeared on the main screen now appears in the small window, and vice versa.

Most of the picture functions, such as freeze and slow motion, operate on the PIP picture. However, only one picture, the main screen, will have audio.

The PIP circuit is normally a plug-in module. The signal processed for the PIP window is normally provided by the A/V switching circuit. However, some manufacturers insert the PIP circuit directly into the video path. The PIP window is only displayed when the user calls for it. *Figure 5-19* shows block diagrams of the two ways PIP can be used with respect to the video. Exactly how the PIP circuits process the signal is manufacturer dependent.

Troubleshooting the PIP

The PIP circuits from the various manufacturers are not the same. When you troubleshoot the PIP circuit in a television, refer to the documentation from the manufacturer or to the PHOTOFACT schematic for that brand and model.

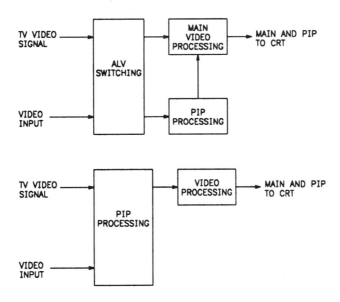

Figure 5-19. A PIP block diagram.

Note: *Some CRTs have been designed differently than described here. However, all of them operate in a similar manner. The one we describe is very common.*

If the PIP circuit is faulty, you may not see color in the small picture window or see a small picture at all. If there is no picture in the small window, check the video signal input waveform at the PIP processing circuits for a proper signal. Then, check the supply voltages to make sure the PIP circuit is receiving the correct voltage.

If there is no color in the small picture, measure the luminance and color signals at the PIP's input stage. Also, check that the chroma signals are being received by the PIP circuit. If the input signals are being received, the PIP circuit is probably faulty.

If the small picture rolls vertically, but the large picture is normal, check the vertical sync signal at the PIP circuit input. Then, check the output of the PIP circuit's sync signal processor. If the small picture is not stable horizontally or takes a while to become stable, check the horizontal and vertical sync waveforms at the PIP processing IC.

How a CRT Works

A television picture tube (CRT) is a glass envelope. The viewing screen is usually rectangular with a width that is 4/3 of the height. The ratio of the horizontal width to the vertical height is called aspect ratio.

The neck of the CRT is between one and two inches in diameter, and contains the electron guns, heater elements, and typically four grids. The neck terminates with a series of pins that connect the CRT to the rest of the chassis.

In addition to the components in the neck, the CRT contains a second anode. The second anode is not connected through the CRT socket, as arcing between the pins could occur due to the extremely high voltage potential. A conductive coating lines the inside of the CRT between the second anode and a high voltage socket commonly called the second anode button. The coating also connects the shadow mask and the phosphor screen. After the electron beam passes through all of the grids, it must pass through the shadow mask. The shadow mask (also referred to as the aperture) is a thin piece of metal with thousands of tiny, vertical slit-shaped openings. The shadow mask is located about half an inch behind the phosphor screen, and is mounted with special bimetal holders. The shadow mask is a final effort to make certain that the beam of electrons hits exactly the right color dots, as fully as possible. (See *Figure 5-20*.) When the CRT is initially fired up, it is relatively cold. As the electrons bombard the shadow mask, the electrons which do not pass through hit the mask and are dissipated as heat. Because of the bimetal holders, as the mask heats up, it moves closer to the phosphor screen to maintain proper

CHAPTER 5: TROUBLESHOOTING VIDEO CIRCUITS

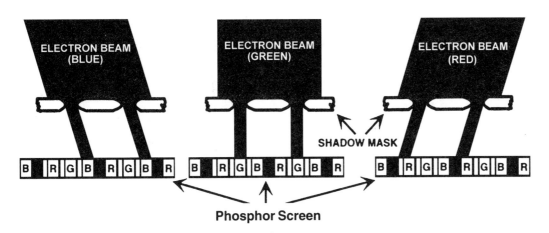

Figure 5-20. A shadow mask.

physical alignment due to the expansion of the slits. This is why it is generally recommended to allow at least a ten-minute warm-up time before any adjustments are performed. The intensity and focus of the individual electronics beams are determined by the grids.

The air inside of the CRT is evacuated to form a vacuum. In a vacuum, heated oxide surfaces emit large quantities of electrons that form a cloud. If an electrode near the oxide surface is made positive, the electrons travel toward it. The part that emits electrons is the cathode, and the grids are the positive attraction.

The control grid (*Figure 5-21*) has a small hole or slot in it. When the electrons from the cathodes are attracted by the positive charge on the control grid, the electrons pass through all of the grids. This is because the electrons are attracted to the more powerful charges of the following anodes. This process of passing the electron beams through the grids narrows the beams. As the beams travel toward the final anode, they accelerate and travel in a straight line down the axis of the CRT toward the center of the screen. When the beams strike the screen, it excites the phosphor which creates a bright spot on the screen. The deflection yoke, controlled by the horizontal and vertical circuits, bend the electron beams across the face of the fluorescent-coated screen in order to form the images.

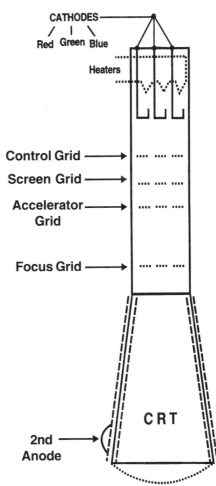

Figure 5-21. A CRT control grid.

Figure 5-22. An external degaussing is needed to clear up this screen.

Refer to the section *Working With a CRT* in Chapter 2 for basic guidelines on working with a CRT.

Troubleshooting a CRT

Most of the symptoms pointing to a defective CRT can also be produced by components other than the CRT. To quickly determine whether the source of the problem is the CRT or other components in the chassis, use a CRT tester to test the CRT. You also can use a color picture test jig to replace the picture tube and see if the picture returns to normal. Either way, you know relatively quickly whether the CRT needs to be replaced.

The following are some symptoms of a defective CRT:
1. Sound, but no picture. In this case, the screen lights but does not produce a picture. If there is a scanning raster, the problem is probably caused by a circuit in the chassis. Also, incorrect voltage on the cathode or grid circuits can cause the CRT to not produce a picture.
2. No scanning raster or an intermittent scanning raster. The screen is entirely black and there is no scanning raster or scanning lines. This might be caused by very low or very high voltage to the CRT. If the voltage is too low, the electron gun does not receive enough voltage to drive the cathode terminals or focus anode terminals. If the voltage is too high, you will see blooming—a washed-out picture with poor contrast or brightness—on the screen.

3. The colors on the screen appear dim or the picture is negative. Three or more of the color guns might have an oxidation buildup. These symptoms also can point to a voltage problem. If the picture appears to be a negative of the intended picture, the problem might be a shorted CRT. If the CRT appears to be working correctly when tested, look at the luminance circuit and the output from the video amplifier circuit as you troubleshoot the chassis.
4. A spot (blotch) appears on the screen, as shown in *Figure 5-22*. This happens if the shadow mask is faulty or the CRT is heavily magnetized, requiring external degaussing.
5. Little control of brightness, focus or contrast. Reduced brightness can be caused by an open heater element or no voltage reaching the heater element. If the CRT is shorted, you cannot control the brightness. There also might be a high voltage problem. If the CRT appears to be working correctly, look at the luminance circuit and the output from the final video amplifier circuit. Reduced focus or intermittent focus problems can be caused by a defective picture tube socket or improper focus voltage. Look at the pins on the CRT socket to see if they are broken, cracked or corroded. When you measure the focus voltage with a high-voltage meter and the voltage is too low, there might be a defective focus control or a leaky spark gap assembly. If the voltage is too high, the CRT, the flyback circuit, or components on the CRT board might be faulty.
6. High-voltage arcing. This can be caused by excessive voltage or by dirty connections. It also can be caused by a crack in the glass of the CRT that lets air in the picture tube. If there are cracks present, the CRT is unusable. If there are no cracks, you can discharge the tube, clean the leads and the tube area, and check the anode for excessively high voltage. You also can replace the high-voltage cable and plug.

If you suspect a high- or low-voltage problem, use a high-voltage probe to test the anode voltage and focus voltage.

Static and Dynamic Convergence

The electron beams in a picture tube come from three different electron guns. These three beams must pass through the same small hole or slot in the CRT's grids. Then, each of the beams—red, green and blue—strikes in the corresponding color phosphor on the picture tube. Also, it is very important that the red beam strike only the red dots in each group of color dots on the screen, the blue beam strike the blue dots, and the green beam strike the green dots.

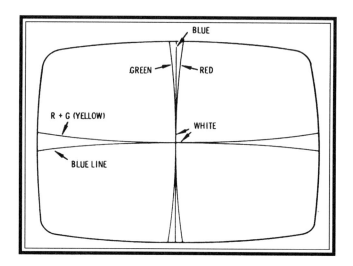

Figure 5-23. A picture with misconvergence.

Regardless of the convergence system used the electron beams are magnetically bent so that they all pass through the small slots in the grids and are projected through the shadow mask to the appropriate colors on the screen. Dynamic convergence is the process that controls where the beams land while the beams are scanning the screen. Static convergence is the meeting of the beams at the screen's center and is made possible by three small magnets that are in the convergence yoke. Static convergence is the starting point for dynamic convergence.

The beams that go to the corners of the screen travel further than the ones that go to the center of the screen. Therefore, those beams must receive additional magnetic corrections in order to land on their appropriate colors—this is accomplished by dynamic convergence.

Convergence Adjustments

If the beams become separated, called misconvergence, the effect worsens as the beams approach the edges of the screen, as shown in *Figure 5-23*. On an otherwise normal signal, dynamic misconvergence appears as color fringes or shadows around the image where the beams do not converge.

Before adjusting convergence, first perform a purity adjustment:
1. Operate the receiver for 15 minutes. Use a degaussing coil to demagnetize the CRT.
2. Turn off the red and blue guns.
3. Loosen the yoke and remove the rubber wedges.
4. Slide the yoke back as far as possible.

CHAPTER 5: TROUBLESHOOTING VIDEO CIRCUITS

5. Move the purity magnets to center the green vertical band.
6. Slowly slide the yokes forward to obtain a uniform green screen.

To adjust the convergence:
1. Tune in a crosshatch pattern.
2. Turn off the green electron gun, so that only the red and blue are operating. Converge the red and blue beams at the center of the screen by rotating the four-pole static convergence magnet on the neck of the CRT. Magnets are shown in *Figure 5-24*.
3. Turn on the green electron gun.
4. Align the red and blue beams with the green at the center of the screen by rotating the six-pole static convergence magnet.
5. Tighten the convergence magnet nut lock. If there is no seal, melt wax and reseal the assembly.
6. Tilt the yoke up or down to converge the vertical lines at the top and bottom of the screen, and the horizontal lines at the left and right sides of the screen.
7. Tilt the yoke right or left to converge the horizontal lines at the top and bottom of the screen, and the vertical lines at the right and left sides of the screen.
8. Repeat the convergence procedures as necessary.
9. Replace the rubber wedges.

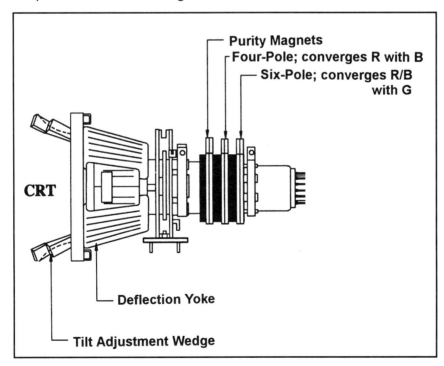

Figure 5-24. A yoke magnet.

Note: In older televisions, the convergence was controlled by three magnets and three coils mounted on the CRT neck. Convergence was performed by a convergence circuit feeding the appropriate waveforms for the three coils. Typically, the waveforms were adjusted by ten controls and two coils in the convergence circuit. Refer to the appropriate PHOTOFACT for proper convergence procedures for this type of circuit.

Quiz

1. What is the last stage of the IF amplifier circuit?
2. At what point can you first measure the composite video signal?
3. What circuit controls where the audio/video signal is going?
4. Which signals do the comb filter separate?
5. Why is a delay line used?

Key

1. A video detector.
2. At the first video amplifier, or the output of the video detector.
3. The A/V switching circuit.
4. Chroma and luminance.
5. To delay the luminance one microsecond from chrominance.

Chapter 6
TROUBLESHOOTING TELEVISION AUDIO

Note: When you check voltages in the troubleshooting procedures, refer to the television's schematic for the correct voltages.

Chapter 6
Troubleshooting Television Audio

The audio circuits in newer televisions are usually contained in one IC. A block diagram outlining the television's audio system is shown in *Figure 6-1*. In a television, the audio signals go through the video IF processing stages prior to arriving at the audio processing circuits. Therefore, problems with the television's sound might not occur in the audio processing stages, but earlier in the television's tuner or IF sections.

Processing Audio Signals

In most televisions, the audio signals are separated from the video signals, color signals, and sync signals at the video detector, which is part of the video IF circuit. Starting at the point where the audio signals are separated from the other signals, the audio signals are processed only as audio. The other signals are handled by other circuits in the chassis. See *Figure 6-2* for a typical schematic of a mono audio processing IC.

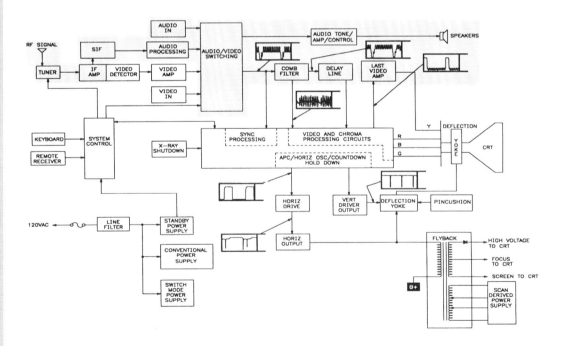

Figure 6-1. The shaded areas show the audio stages.

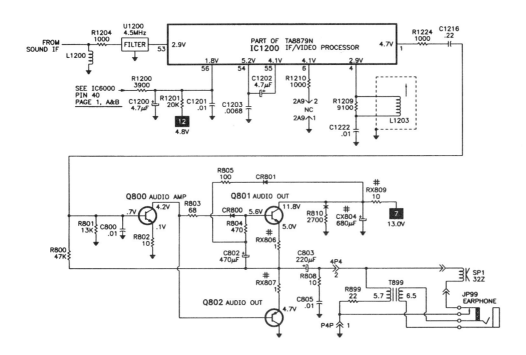

Figure 6-2. A schematic of an audio processing IC.

Sound IF Amplifier

The video detector outputs the FM audio signals to the audio IF stage where the 4.5 MHz audio IF signal is created by beating together the sound and video frequency (41.25 mixer frequency and the 45.75 oscillator frequency). The audio signals, at this point, are really FM radio signals. Remember, the video signals are carried on AM frequencies, and the audio signals are carried on FM frequencies.

The sound IF amplifier is a fixed radio frequency amplifier—it amplifies only one frequency, 4.5 MHz. There can be slight variance in the signal frequency, but the signal must be very close to 4.5 MHz for the audio processing stages to produce the correct sound. The 4.5 MHz signal is amplified and output to the audio IF detector.

When you troubleshoot the audio IF stage, measure the IC terminal voltages and compare the values you get to the values on the schematic.

Audio IF Detector

The audio IF detector receives the 4.5 MHz FM signal and traps the clean audio signal from the FM signal. The audio detector rarely fails. However, if it is a tunable detector that has not been tuned properly, you might hear a buzz in the sound. When the signal is clean, it is output from the detector and sent to the audio amplifiers, volume/tone control circuits, and audio output circuits.

Stereo

When an FM audio signal is transmitted, the signal contains the combined left and right signals (L+R). There also is a modulated signal which is the left signal minus right signal (L-R), which is necessary for the stereo sound. This signal is carried on an AM frequency carrier that is differentiated by the signal strength from the video signals also carried on the AM frequency. This encoding signal is used by the television to demodulate (decode) the stereo signal and produce stereo sound. Another part of the regular audio signal, called a pilot signal, tells the receiver whether or not the signal being received is stereo. Remember, many television stations still do not broadcast in stereo. Sometimes there is also a SAP (second audio program) signal which is part of the regular audio signal. SAP is used to broadcast in a second language. All of these signals are considered to be composite audio signals.

Non-stereo televisions can receive all of these signals, but if the television does not have a stereo demodulator, the television cannot decode the stereo signal. The 15.734 kHz FM modulated signal goes to the audio processing circuits, where it is demodulated, filtered and amplified. It then goes to either the A/V switching circuit or the audio amplifier/control circuits. In addition, some televisions can produce a stereo-like sound that is not true stereo. This is done by splitting the audio signal between two speakers. Likewise, some televisions can receive but not demodulate and reproduce true Dolby™ Surround Sound.

The pilot signal is used for detecting stereo broadcast signals. After the pilot signal is detected, the audio processing circuits tell the system control IC that stereo is available. The system control IC will determine if the audio signal is a stereo, SAP or mono signal. The audio processing circuit then outputs the proper signal to the audio amplifier/control circuits.

Low bandpass filters are used to separate the composite audio signals. The 15.73 kHz regular monaural audio (L+R) signal is sent to an amplifier where it is amplified and output to the audio matrix.

The 31.468 kHz AM modulated stereo signal (L-R), and the 15.734 kHz pilot signal, are sent to the audio processing circuits. The 31.468 kHz is demodulated into the L-R stereo signal. Then, the demodulated stereo signal is sent to the noise reduction circuit where the noise resulting from the high frequency of the pilot signal is eliminated. The cleaner L-R signal is sent to the audio matrix, where it is combined with the regular audio signal (L+R) in order to produce stereo sound.

The SAP signal, which is filtered out by the 78.670 kHz filter, also passes through a demodulator, then through a noise reduction circuit because it carries noisy high frequencies. The SAP signal then is on standby until the SAP function is activated by user choice.

Surround Sound

Surround Sound creates in a television the effect of movie theater sound. Three common types of surround sound circuits are:
1. Matrix is stereo sound in which one speaker produces the regular stereo sound and the other speaker produces the L-R signal. The result is a richer sound.
2. Hall matrix is the same as matrix except that there is a time delay of 32 msec added to the rear speakers so that the sound has some reverberation and creates the effect of being in a concert hall.
3. Dolby™ reproduces sound recorded in Dolby™ mode. There is a left and right channel. Also, there is a center channel, which is the monaural signal (L+R). There is a 20 msec time delay which causes the listener to think that the sound is coming from all over the room. This type of system requires a center speaker. Dolby ProLogic™ further enhances the total sound effect and provides sound from four speakers—left, right, center and surround.

If the television program or movie you are watching is recorded and transmitted using true Dolby™ Surround or Dolby ProLogic™, you will need Dolby™ processing in your television or in an external surround sound processing system attached to your television.

Audio Amplifier

The audio amplifier amplifies the clean audio signal. The volume control is in the audio amplifier. When the signals are sufficiently amplified, the signals from the audio amplifier are sent to the speaker(s) which produce the sound. Normally, the audio amplifier is equipped to mute the sound whenever the system control tells it.

Troubleshooting Symptoms

Listen carefully to the sound from the speakers. As you do, vary the volume control. Usually, you can hear problems with a television's audio system. The sound might be missing, weak, distorted, intermittent, or contain interference. However, problems with the sound system can begin outside of the audio circuit. If the sound problem is accompanied by a picture problem, the problem is probably being caused by a faulty IF processing circuit. For example:

1. Poor sound quality accompanied by an unstable picture suggest a problem in the AGC circuits.
2. Sound and picture do not track properly, suggesting a problem with the detector coil or poor sound alignment.
3. A hum or buzz in the sound accompanied by a running sync suggest a problem in the sync or AGC circuits.
4. Squealing or whistling, distorted or intermittent sound, or no sound, even at maximum volume, accompanied by a normal picture and raster suggest a problem in the audio processing circuits.

To troubleshoot the audio processing circuits, you can use an MTS signal generator, and an oscilloscope to inject a signal beginning at the output of the video detector and test each output to input in the audio circuits until you reach the speakers. Also, you can use an external audio amplifier and trace the signal from the input of the last amplifier back through the audio circuits until you reach the output of the audio processor.

Note: In televisions with remote-controlled volume, the volume level is determined by the system control signals that are sent to the audio circuits. See Chapter 10.

No Stereo

If there is no sound at all, check all of the audio processing circuits. If the television is supposed to receive stereo and the sound is not in stereo, make sure the program you are listening to is broadcast in stereo. If the program is supposed to be in stereo, and the monaural sound is not distorted, the pilot demodulation circuit (L-R) might be faulty. Using an oscilloscope, check that the correct signal is being input into the demodulator. Remember, the signal is really the composite audio signal minus the SAP signal, or in other words, all signals below 45 kHz. Then, test the output from the demodulator circuit. Continue testing circuits, and the components and connections around the circuits in the stereo demodulator path, until you locate the faulty circuit, component or connection. If you inject a test signal, remember that to test the regular audio processing circuit, use a mono signal of 15.734 kHz, FM modulated. To test the stereo (L-R) signal, use a signal of 31.468 kHz, AM modulated.

If there is no monaural sound but there is distorted stereo sound, check the audio IF processing IC.

If there is noise in the stereo signal, check all of the circuits in the stereo processing IC, starting with the low bandpass filters. The filters might be letting in extraneous frequencies that the noise reduction circuits are not eliminating. Or, the noise reduction circuits might not be eliminating the high frequencies correctly, which might indicate a faulty IC.

No SAP

Follow most of the troubleshooting guidelines for troubleshooting stereo. Also, if you inject a test signal when troubleshooting the SAP processing circuits, use a 78.670 kHz FM modulated signal.

Troubleshooting Example

If you suspect that a speaker is faulty or open, inject an audio signal into the input of the speaker and listen. If you hear a hum or tone from the speakers, the speakers are probably operating normally. However, you still have to listen carefully to the audio produced by the speakers. There could be a warped, torn or broken place in the cone. Also, measure the resistance in the speaker's coil using an ohmmeter. The number of ohms should match the value on the speaker's specifications. Also, if the speaker is working, you should hear a single click as you attach the ohmmeter.

If you do not hear a tone, the speaker may be open:
1. Substitute another speaker for the suspected speaker.
2. Turn the volume control to midrange, because you cannot conduct tone tests with the volume turned down.
3. Check the audio amplifier's voltages. The values should match the values on the speaker schematic. If they do not, there might be an open connection or a leaky transistor or capacitor in or around the audio amplifier stage. Also, the voltage being received might be too high or too low. Use signal injection to check the inputs and outputs, and locate the faulty component, connection, or IC.
4. Test the components around the audio amplifier stage for opens, shorts or value changes. The values should match the values on the audio schematic. If the values are not correct, replace the faulty component.

No Sound

You can use signal tracing with an audio signal tracer to test the circuits when there is no sound. In audio stages, where there is sufficient gain, you can use a 500uF electrolytic capacitor and a speaker with a clip wire. Clip one side of the speaker to a good ground and the other speaker lead to the capacitor. Input a signal from an audio signal generator, and using the end of the capacitor lead, probe for the signal at the output.

You can use a voltmeter to test the voltages, and an audio signal generator and an oscilloscope to check the waveforms:
1. If the speakers do not produce sound, not even a hissing noise when the volume control is at maximum, there might be an open connection in or around the audio amplifier.
2. If there is a humming sound, the speakers and audio power amplifier are working correctly, the problem is probably with the voltage from the power supply or the audio driver amplifier.
3. If you do not hear hum, the problem is with the speaker(s) or the audio power amplifier. For example, the speaker could have a broken wire in the voice coil or an open coupling capacitor to the speaker.

Weak Sound

Weak sound can be caused by a leaky or open transistor or capacitor, a resistance change or voltage that is too high or too low. Also, if the cone in the speaker is defective or frozen, the sound will be muffled, distorted or weak. If the sound is weak and slightly distorted, some frequency variations may be getting through the detector coil. Therefore, adjust the detector coil to fine tune the audio take off if the circuit is adjustable.

If the detector coil and the speaker cone are working properly, use signal tracing to test the circuits. Supply the signal with a signal generator and trace the signal with an oscilloscope. Also, measure the voltages on all IC pins and for all components around the IC to make sure the voltages supplied to the IC are correct. When checking components, you may need to lift one side of a component to break the circuit before you test it for an accurate measurement.

Also, a change in the resistance of the collector and emitter resistors can cause the sound to be weak or distorted. Use a DMM to check for open connections or leaky components. Also, make sure to check the capacitors.

CHAPTER 6: TROUBLESHOOTING TELEVISION AUDIO

Squealing or Whistling Sounds

If you hear a squealing or whistling sound, the source of the problem can be in other circuits of the chassis, but usually there is a faulty transistor or capacitor in one of the audio stages. Check each of the components around the IC, and compare the measurements with those on the schematic.

Humming and Putt-Putting Sounds

If you hear a humming sound, there might be a faulty capacitor or filter regulator in the circuit. Also, there might be a short or a bad solder connection. Listen for the hum with the volume turned up and down. Use signal tracing and test the voltages at all input and output connections. Check the IC and all of the components around the IC. A poor ground connection or broken path in the audio circuits can also produce hum or putt-putting.

Most audio problems are found to be in the audio circuits, but putt-putt sounds, also called "motor boating", can be caused by wiring problems in the horizontal circuits. If the picture is also erratic, check the other video circuits as well. When you locate a transistor that might be the source of the problem, short the emitter and base terminals of the transistor. If the motor boat sound disappears, replace the faulty transistor.

Crackling or Popping Sound

If you hear a crackling or popping sound when you turn on the television, check the audio amplifier stages and the power supply filters. Faulty transistors will crackle when heating up. Therefore, you can use a cold spray to test the transistor. Take the transistors out of the circuit before you test their resistance. Open filter capacitors in the audio circuits may also cause popping sounds.

Distorted or Intermittent Sound

If the sound is distorted, suspect a faulty speaker, transistor, resistor, sound alignment or IC. These components cause most of the sound distortions you hear from the speakers:
1. If the sound is tinny or "mushy" when the volume is turned down, or dull, low-pitched, and vibrates (often called blatting) when the volume is turned up, check for a faulty transistor, resistor, speaker, sound alignment, or IC.
2. If the sound is extremely distorted, check the speaker, the sound alignment, and check for leaky or shorted transistors, capacitors or

connectors. Remember to test the bias and base resistors when you remove a transistor from the circuit for testing.

If the sound is intermittent or erratic as if the audio frequency is drifting, or if the sound is muffled, the detector coil might need adjustment, or the audio IF stages may be defective. Use a special adjustment tool and make small adjustments to the coil until the sound is clear. Use an oscilloscope to test the audio signal's waveform, output from the detector coil, until the signal is adjusted properly.

Loose wiring, and open or shorted connections can also cause intermittent sound problems. With the chassis and speakers on, lightly pull on components to see if they are loose. With the chassis and speakers on, use a cold spray to locate cold solder joints or heat-sensitive components.

Quiz

1. What is the SIF frequency?
2. Describe a key purpose of the pilot signal regarding stereo sound.
3. Is the SAP signal AM or FM modulated?
4. At what stage is the volume control located?
5. At what frequency is the stereo signal modulated?

Key

1. 4.5 MHz.
2. It tells the system control that a stereo signal is present.
3. FM.
4. The audio amplifier.
5. 31.468 kHz.

Chapter 7
TROUBLESHOOTING DEFLECTION CIRCUITS

Note: When checking voltages in the troubleshooting procedures, refer to the PHOTOFACT schematic from Howard W. Sams for the television's brand and model for the correct voltages. Also, make sure to check all supply voltages.

Chapter 7
Troubleshooting Deflection Circuits

The horizontal and vertical deflection circuits control the horizontal and vertical scanning that produce the picture on the television screen. In other words, the deflection circuits cause the electron beams to move in a manner that produces a scanning raster on the screen:

1. The sync separator divides the sync signals into horizontal signals and vertical signals, removes extraneous signals, such as video signals, from the sync signals, amplifies the signals, and outputs the signals to the appropriate deflection circuits.
2. The horizontal deflection circuit moves the electron beams from the left to the right on the screen, then back to the left 15,734 times every second. The horizontal scan operates at a frequency of 15.734 kHz.

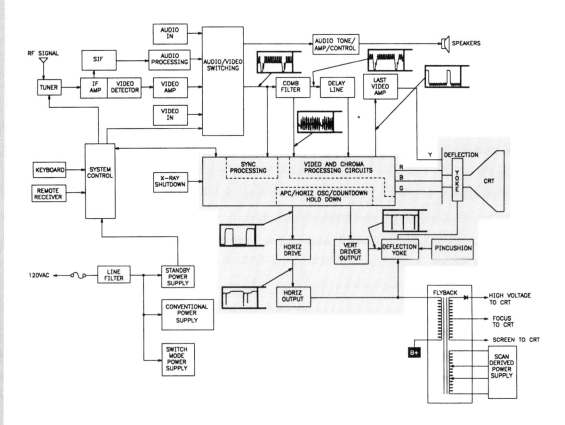

Figure 7-1. The deflection circuits (shaded).

3. The vertical deflection circuit moves the electron beams from top to the bottom of the screen, then back to the top 60 times every second. The vertical scan operates at a frequency of 60 Hz.

A basic deflection circuit diagram is shown in *Figure 7-1*.

The horizontal output circuit also produces the high voltage needed to power the CRT and other circuits in the chassis. This use of the horizontal output circuit is described in Chapter 4, **Troubleshooting Power Supplies** and Chapter 8, **Troubleshooting High-Voltage Circuits**.

If you suspect that an IC is faulty, gently press the IC to make sure the pins are correctly seated if the IC is mounted in a plug-in socket. If the problem persists, continue with the troubleshooting procedures. Measure voltages on all pins. Also, check the solder connections on all pins.

The configurations of ICs and components that make up the deflection circuits can vary according to the manufacturer. Refer to the PHOTOFACT schematic for the television's brand and model for the exact composition of the circuits and component values.

Sync Separator

When the television station transmits the composite video signal, shown in *Figure 7-2*, it contains the video signal, the audio signal, the color signal and the sync signal. The sync signal synchronizes the horizontal and vertical scanning on the CRT screen with the original signal broadcast from the station.

The video processing circuits separate the sync signals from the video signals. The sync separator circuit, like the one shown in *Figure 7-3*, filters out any extraneous signals, including video and noise. The clean sync signals are output to low-pass filters that separate the 60 Hz vertical signals from the 15.750 kHz horizontal signals.

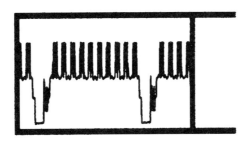

Figure 7-2. A composite video signal.

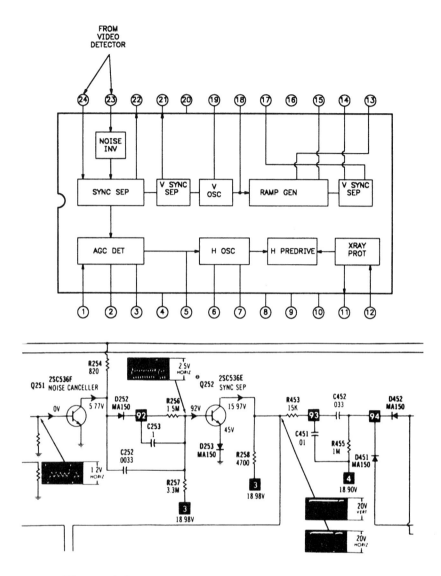

Figure 7-3. A sync separator circuit, internal to IC.

The first stage in the sync separator separates the sync signals from the video signal and any noise that remains in the signal. Then, the vertical integrator and the sync amplifier select the vertical sync signal from the total sync signal, amplifies the vertical sync signal, and sends it to the vertical deflection circuits. In the same manner, the horizontal differentiator selects the horizontal sync signal, amplifies the horizontal sync signal, and sends it to the horizontal deflection circuits.

Troubleshooting the Sync Separator

Picture problems, like those shown in *Figure 7-4*, suggest that the sync separator is not working properly.

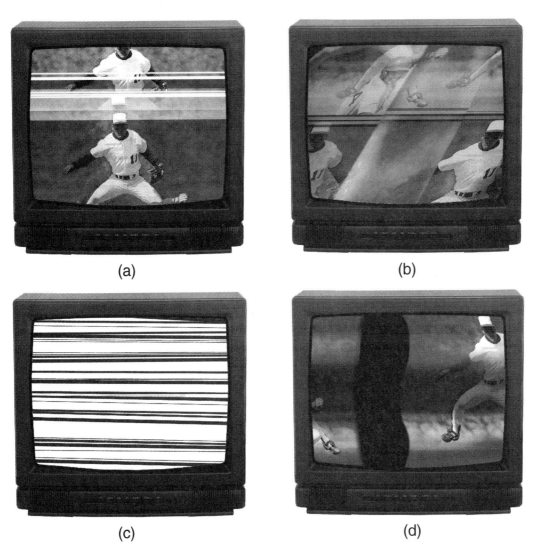

Figure 7-4. Sync signal problems: moderate loss of vertical sync (a); moderate loss of horizontal sync (b); severe loss of horizontal sync (c); loss of horizontal and vertical sync (d).

Loss of vertical sync causes the picture to roll up and roll down, and not lock vertically, even when you adjust the vertical hold. This problem suggests that the vertical integrator/amplifier stage is faulty and cannot maintain the vertical hold. If the vertical deflection circuits were faulty, the picture would roll either up or down, but not in both directions. Use an oscilloscope to check the waveform from the output of the vertical integrator/amplifier stage. If there are horizontal signals or noise in the waveform, the vertical integrator/amplifier is not selecting only the vertical signals.

If you see horizontal tearing or just horizontal scan lines, try to adjust the horizontal hold. If the horizontal hold control does not lock the picture horizontally, check the horizontal differentiator/amplifier or the APC (automatic phase control) circuit, described later in this chapter. Use an oscilloscope to check the waveform from the output of the horizontal differentiator/amplifier stage. If there are vertical signals or noise in the waveform, the horizontal differentiator/amplifier is not selecting only the horizontal signals. See the section on the APC, later in this chapter, for more information on troubleshooting the APC.

A problem with multiple symptoms, such as horizontal tearing and vertical rolling, and the blanking bar appearing on the screen, indicates a problem before the sync separator circuit. Look carefully at the waveform output from the video detector circuit. If the video detector output signal is correct, trace the signal through the sync separator. If the video detector output signal is not correct, measure the outputs and inputs for all circuits that process the sync signal. Most sync problems in this area occur at or after the video the detector. IF stages rarely cause this type of problem.

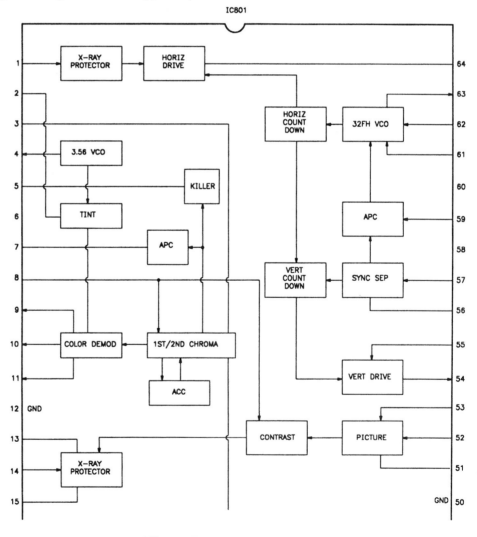

Figure 7-5. An APC circuit.

CHAPTER 7: TROUBLESHOOTING DEFLECTION CIRCUITS

As you are troubleshooting the sync circuit IC, check the supply voltages to the IC, and check all components and connections around the IC. If the source of the problem is the sync separator circuit IC, replace the IC.

Automatic Phase Control (APC)

The APC circuit, shown in *Figure 7-5*, is part of the horizontal deflection IC and keeps the horizontal oscillator synchronized with the sync signal sent from the television station. This circuit rarely is deflective.

There are two inputs into the APC. One input is the sync signal pulse from the sync separator. The other input is a reference signal from the horizontal oscillator from any circuit past the horizontal oscillator. The APC compares the horizontal sync pulse and the reference signal for a phase difference. If it detects a difference, the APC produces a correction voltage. The correction is made when the polarity of the correction voltage adds to or subtracts from the horizontal oscillator's normal bias. The correction output voltage varies according to how close the two inputs are. The correction voltage is output to the horizontal oscillator, and the horizontal oscillator changes frequency accordingly. In some televisions, the APC does not output directly to the horizontal oscillator. Instead, there is a reactance stage between the APC and the horizontal oscillator, as shown in *Figure 7-6*.

Troubleshooting the APC

You can check the APC by adjusting the television's horizontal hold control. If the control adjusts the voltage and changes the picture, the APC is working at some level. You can connect an oscilloscope to the output of the APC to see the waveforms as they change according to the adjustments you make. However, if you see horizontal tearing or weaving, or a jittery picture, the APC may not be working correctly.

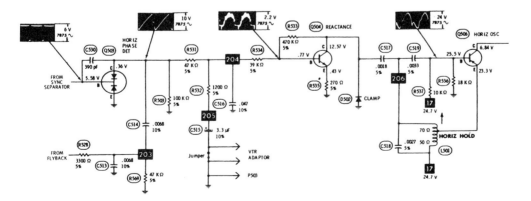

Figure 7-6. A reactance stage circuit.

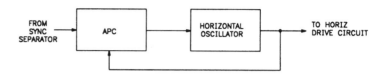

Figure 7-7. APC and the horizontal oscillator form a closed loop.

The APC and the horizontal oscillator form a closed loop—a series of circuits whose outputs affect their inputs, and whose inputs affect their outputs. This loop is shown in *Figure 7-7*. Also, the inputs and outputs originate outside the closed loop. This can complicate the troubleshooting process. Therefore, when you troubleshoot a jittery picture, measure the inputs and output from the APC first. Then, measure the input and output from the horizontal oscillator. If both are operating properly, expand your troubleshooting to include the circuits that originate the inputs to the loop.

You can quickly test the APC by jumping the circuit and using a voltage generator to substitute for the input voltages. Then, you can vary the voltages and measure the outputs. If the APC is working correctly, use signal tracing or signal injection to test the inputs and outputs of the circuits that affect the APC circuit. Also, make sure you check the components and connections around the IC, including transistors and diodes.

If you determine that the APC is faulty, replace the horizontal deflection IC.

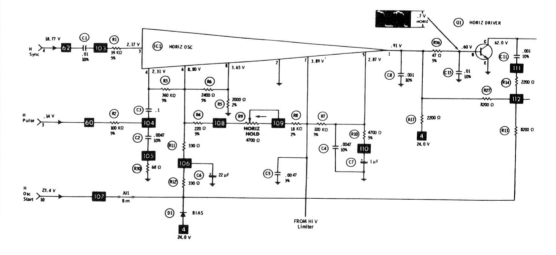

Figure 7-8. A horizontal oscillator circuit.

Horizontal Oscillator

The horizontal oscillator circuit, shown in *Figure 7-8*, produces 31.5 kHz signals that maintain the picture's horizontal stability. The horizontal oscillator also produces the drive signal for the horizontal output circuit. The horizontal hold control adjusts the frequency of the oscillator which, also controls the output signal as well as the APC reference signal.

The horizontal oscillator is an amplifier that has a feedback to the APC. If the feedback is not in phase with the input when the two signals are compared in the APC, the APC generates a correction voltage. The horizontal oscillator can be many types; however, regardless of the type, the purpose is the same:
1. Countdown circuit.
2. Push-pull amplifier.
3. SCR.

The input to the timing circuit is the signal from the horizontal oscillator, the horizontal sync pulse and the correction voltage from the APC. The output from the timing circuit is a 15,734 Hz signal that drives the horizontal output circuit.

Troubleshooting the Horizontal Oscillator

If you see a picture on the screen, the horizontal is working, even though it might not be working correctly. If the oscillating output frequency is too low, the oscillator is running too slowly and the picture on the screen will tear downward to the left. If the oscillating output frequency is too high, the oscillator is running too fast, and the picture on the screen will tear upward to the left.

You can quickly check the operation of the horizontal oscillator by adjusting the horizontal hold control. If you cannot adjust the picture using the horizontal hold control, the horizontal oscillator or the APC might be faulty.

If you suspect that the feedback loop is not working properly, measure the output from the horizontal oscillator. Then, compare the output waveform you get to the output waveform on the PHOTOFACT schematic. If the output waveform is correct, measure the supply voltages to the oscillator.

If the supply voltage is correct, check the APC as described earlier in this chapter. Then, measure the inputs and outputs of the components that are around the IC and the circuits that impact the voltages in the horizontal deflection circuit. Also, check the soldered connections.

Note: In newer televisions, the oscillator runs at 503 kHz and is passed through countdown circuits to arrive at the horizontal frequency. The frequency of the oscillator must be a derivative of 15.734 kHz; however, the function and operation is the same as the example (Figure 7-8).

Check the base for the correct waveform. If the input waveform to the horizontal oscillator is correct but the output from the oscillator is not correct, the horizontal oscillator circuit may be faulty.

If the frequency output from the horizontal oscillator is too low or too high, another component may have increased or decreased its frequency, and the time constant has changed. If the time constant has decreased, look for a faulty resistor, shorted capacitor or a bad coil. If the time constant has increased, look for a faulty resistor or an open capacitor.

Shorts can cause erratic responses from the horizontal oscillator. However, if the television is producing no picture and no sound, and you want to check the horizontal oscillator, inject a signal from a signal generator into the input of the horizontal oscillator and supply power to the horizontal oscillator. Then, use an oscilloscope to check the output. This quick test will tell you whether the oscillator is working, even though it still might be faulty. It also tells you that the horizontal oscillator may not be the reason the television is inoperative.

Horizontal Deflection Circuits

Note: The best way to troubleshoot deflection circuits is by signal injection.

The horizontal deflection (drive) circuits, shown in *Figure 7-9*, process the horizontal signal to ensure that the electron beam deflection on the picture tube is centered, and the video signal is correctly synchronized with the original signal broadcast by the television station.

If you suspect that the horizontal deflection IC is faulty, check each of its stages. When you troubleshoot the horizontal deflection circuits, use an oscilloscope to look at the signal waveforms. Use a DMM to check the input and output voltages, and a high-voltage probe to check the focus and the CRT high voltage.

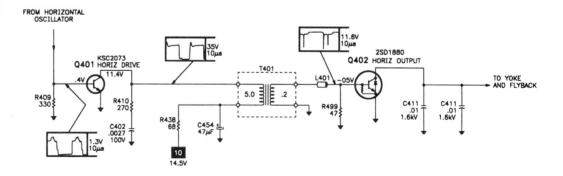

Figure 7-9. The horizontal deflection (drive) circuit.

CHAPTER 7: TROUBLESHOOTING DEFLECTION CIRCUITS

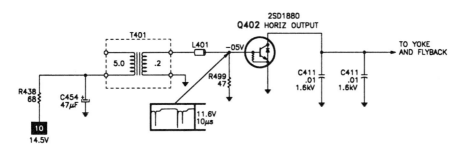

Figure 7-10. A horizontal output circuit.

Horizontal Output Circuit

The horizontal output circuit, shown in *Figure 7-10*, is a transistor that is used as a switching circuit. It performs two functions:
1. It controls the horizontal scan that produces the horizontal lines in a television's picture.
2. It produces high voltage power and low voltage supplies for other circuits and for the CRT. The high-voltage use is described in Chapter 4 and Chapter 8.

The horizontal output circuit is considered a switching circuit because it controls the power to the CRT's deflection yoke and to the flyback. In newer televisions, the damper diode that acts as a switching device is built into the transistor. In older televisions, the damper diode is a separate component.

The output circuit can be directly coupled to the deflection yoke. However, the output circuit can also use a resistor, capacitor, transformer or transistor to couple the signal to the deflection yoke.

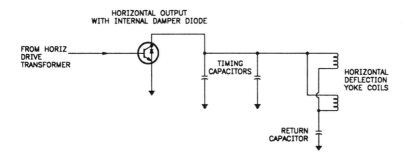

Figure 7-11. A simplified horizontal output circuit with the deflection yoke connected.

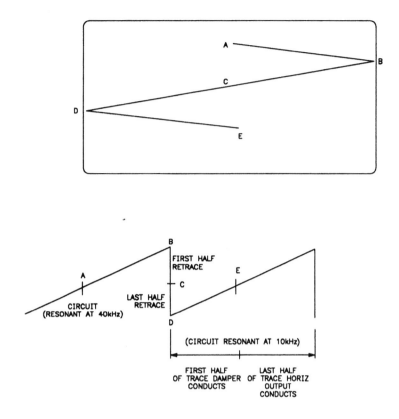

Figure 7-12. One scan line and one retrace sequence.

Horizontal Scanning

The horizontal output circuit is resonant (can be tuned) to 10 kHz. However, in order to rapidly retrace (move back to the left side of the television's screen) when the end of a scan line is reached, other components are added to the circuit so that the circuit is resonant at 40 kHz. These components are the damper diodes, the deflection yoke coil and capacitors, and the timing capacitors. A simplified horizontal output circuit with the deflection yoke connected is shown in *Figure 7-11*.

In older televisions, the damper diode was external to the transistor. The damper diode is connected from the horizontal output transistor's emitter to its collector. The deflection yoke coil is connected to the collector of the horizontal output transistor. The return capacitors are in series with the yoke to ground. For efficiency, the timing capacitor is parallel with the damper diode and the horizontal output transistor. This structure keeps the voltage discharge smaller and linear during a scan sequence—less than the capacitor's total voltage.

CHAPTER 7: TROUBLESHOOTING DEFLECTION CIRCUITS

Figure 7-12 shows one scan line and one retrace sequence. The retrace requires a pulse of 10 kHz. The scan requires a pulse 40 kHz. In order to produce one scan cycle, the voltage must vary from 10 kHz to 40 kHz in an orderly, timed manner.

The scan takes exactly 51 msec. The retrace takes exactly 12.5 msec. In order for the scan to take place in 51 msec, the resonant frequency must be 40 kHz. The retrace requires 10 kHz.

For example, to begin the scan and move the electron beams from the center of the screen to the right side of the screen (A to B in the diagram), there must be a linear increase in magnetic energy on the right side of the screen to attract the electron beams. Therefore, the current in the deflection yoke is increased on the right side, which causes a magnetic field to build up around the right side of the deflection yoke. The magnetic field then attracts the electron beam to the right side of the screen.

When the resonant frequency increases to 40 kHz, the magnetic field around the yoke drops and the magnetic field collapses. At the same time, the current in the deflection yoke starts to increase on the left side, which causes the rapid magnetic field buildup on the left side of the screen. This causes the electron beams to return to the center (B to C) then deflect to the left (C to D). Finally, when the magnetic field on the left collapses, and the field starts to increase on the right side, the beams return to center (D to E) and deflect to the right again. *Figure 7-13* shows the shape of a horizontal waveform.

Triggering a Scan/Retrace Sequence

Refer to the PHOTOFACT schematic for the television you are troubleshooting for the type of triggering device used in the chassis and the voltage values to expect when testing the inputs and outputs.

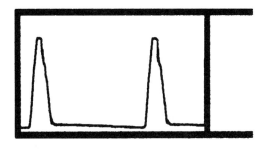

Figure 7-13. A horizontal waveform.

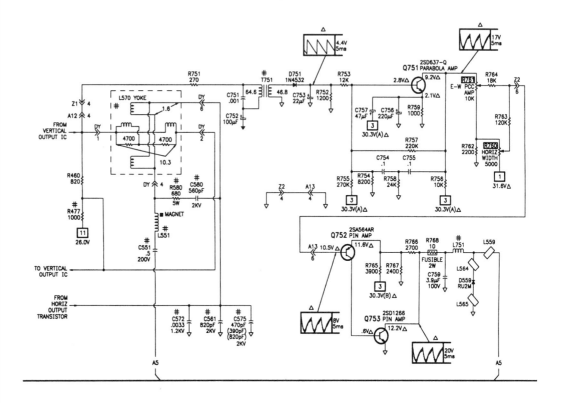

Figure 7-14. A pincushion circuit.

A damper diode acts a switch that causes the transfer of current from the yoke to the capacitor, and causes the scan on the left side of the screen. The horizontal output transistor acts as a switch that causes the transfer of current from the capacitor to the yoke, and causes the scan on the right side of the screen.

An SCR (silicon control rectifier) also can be used to trigger a scan/retrace sequence. The SCR is a switching component that is turned on by a signal pulse. When voltage reverses in polarity or drops significantly, the SCR turns off. When it receives another signal pulse, it turns on again. There usually are two SCRs in a horizontal output circuit. One is used to trigger a scan; the other is used to trigger retrace.

Pincushion Correction Circuit

The pincushion correction circuit, shown in *Figure 7-14*, prevents a distortion called pincushioning, shown in *Figure 7-15*, where the picture appears to pull in at the center and out at the corners. The pincushion circuit increases the voltage to the mid-screen section of the yoke to compensate for the voltage reduction that caused the distortion.

Figure 7-15. Picture pincushioning.

Troubleshooting a Horizontal Output Circuit

If the horizontal output circuit is faulty, the symptoms can vary. If the television is totally inoperative (no sound or picture), the scan-derived power supply from the horizontal output circuit might have a shorted diode. Also, the boost circuit might be bad. Refer to Chapter 4, **Troubleshooting Power Supplies**, and Chapter 8, **Troubleshooting High-Voltage Circuits**, for more information on the scan derived power supply.

If the focus is bad, the picture is dim, the picture brightness is not right, or you see blooming on the screen, there is probably an overloaded circuit. Also, you might hear a ringing sound associated with the picture problem.

If you see a picture width problem, or if the picture folds or weaves, the source is probably a drop in the voltage pulses to the deflection yoke coil. If you see the keystone effect where the picture is narrower in one direction than the other, there is probably a short in a deflection coil.

If you turn on the television and the B+ fuse blows, check for a shorted horizontal output transistor or damper diode. A shorted transistor or diode will cause the fuse to open.

Note: *If the emitter collector junction measures low resistance when checking a horizontal output transistor, be aware of any internal components that may have a low resistance measurement, such as a damper diode or resistor.*

Note: *The best way to troubleshoot horizontal deflection is to inject a horizontal drive pulse into the base of the horizontal output transistor. Problems in the transistor, flyback or yoke can then be detected.*

Note: *Voltage on the collector of the horizontal output transistor can cause damage to many oscilloscopes and voltmeters, due to the high peak-to-peak RF waveform.*

A shorted damper diode or SCR suggests that there may be a high B+ voltage condition. An overloaded circuit can cause shorts. If you replace a damper diode or SCR, make sure you use exact replacements. In newer televisions, the damper diode is part of the transistor.

When you begin troubleshooting, use an oscilloscope and test the input signal to make sure it is correct. If it is not, trace it back through the path of the horizontal oscillator, sync and drive processing circuits until you find the source of the problem. If the input signal is correct, test the output signals and the output voltage from the horizontal output circuit.

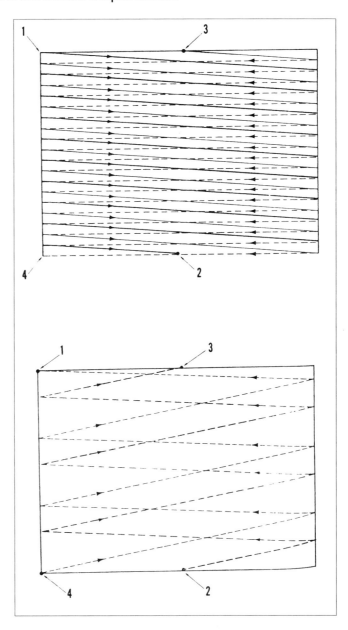

Figure 7-16. Interlaced scanning.

Vertical Deflection Circuits

When one set of 262+ horizontal lines are scanned, the vertical retrace causes the electron beams to be deflected back to the top left of the screen. Then, the next set of 262+ horizontal lines are scanned. *Figure 7-16* shows interlaced scanning and the signals for retrace on the alternate fields.

In the figure, the first field starts at (1) and ends at (2). The second field starts at (3) and ends at (4). During the vertical retrace when the picture is blanked, the beam moves back to (1). Then, the cycle begins again.

Horizontal scanning occurs 15,750 times every second. Vertical scanning occurs 60 times every second. The vertical deflection circuit also supplies a blanking signal so that the CRT's screen blanks during the vertical retrace. A problem with the vertical deflection circuits can cause the picture to roll vertically.

In newer televisions, the vertical deflection circuit may be one or two ICs. The main stages in the IC are the vertical oscillator and the vertical output circuits. However, there may be other support circuits around the main stages, such as an amplifier between the vertical oscillator and the vertical output circuits.

Vertical Oscillator

Vertical oscillators produce a signal that turns off the vertical output transistors. Turning off the vertical output transistors causes the magnetic field in the vertical deflection yoke to collapse. The result is a retrace to the top of the screen. During the retrace, a blanking signal is sent to the vertical output circuits to turn off the scan and blank the screen.

The vertical oscillator operates at 60 Hz and is controlled by the sync pulses it receives from the sync separator. The vertical oscillator is always on, running at a slightly lower frequency than the sync signal from the sync separator. When the sync signal is input, the transistors conduct the signals. When the signal is

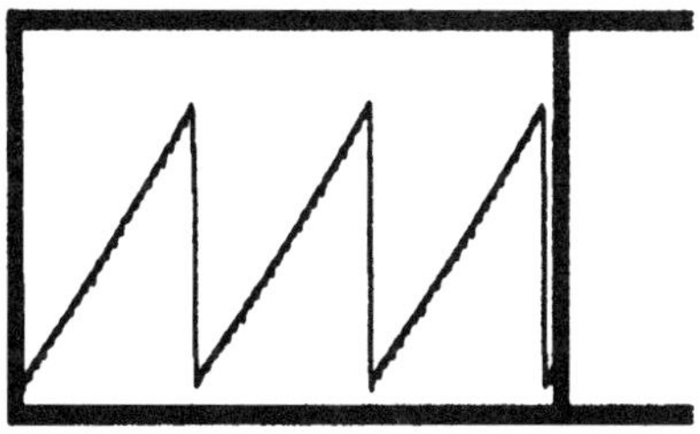

Note: In some televisions, the vertical oscillator is part of a countdown circuit of a horizontal oscillator normally internal to an IC.

Figure 7-17. A sawtooth waveform.

discharged, the frequency drops and once again the oscillator is running at a lower frequency than the sync signal. This pattern of conduction and discharge creates a sawtooth waveform, like the one shown in *Figure 7-17*. Also, since the vertical scan occurs at a much slower rate than the horizontal scan, pulses are sent out between each vertical scan/retrace cycle to keep the horizontal oscillator synchronized with the vertical retrace.

The purpose of the vertical scanning waveform is to trigger the vertical oscillator, to blank the screen during retrace, to keep the horizontal scanning in sync during vertical retrace, and to make sure that alternate fields are interlaced.

Output from the vertical oscillator is a linear signal of 1V to 2V. The output is a modified sawtooth scan signal, shown in *Figure 7-18* which is sent through an amplifier to the vertical output circuit. The amplitude of the sawtooth signal corresponds to the vertical size of the picture screen. The amplifier receives feedback from the vertical deflection yoke and adjusts the signal as needed. The signal cannot be linear if the sawtooth waveform is not linear. If the signal is not linear, the picture is scanned at an uneven rate and will be distorted.

Several types of oscillators can be used in the vertical circuits, such as:
1. Countdown circuit.
2. Sawtooth generator.

The vertical timing circuit, like the countdown circuit shown in *Figure 7-19*, contains several flip-flop circuits that correctly divide the horizontal pulse, and "count down" from the horizontal rate to the vertical rate. The purpose of the timing circuit is to cause the capacitor to discharge at the right time to trigger the vertical retrace. Refer to the PHOTOFACT schematic for the correct input and output values for the television you are servicing.

If you suspect that the timing circuit is faulty, check the supply voltages and the feedback, as well as the components and connections around the IC. Also, you can temporarily replace the IC with one that you know is working properly. If the normal operation resumes, the IC is defective.

Figure 7-18. A modified sawtooth waveform.

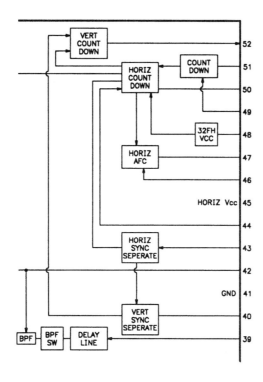

Figure 7-19. A vertical countdown circuit.

Feedback in Vertical Circuits

When a picture is not linear, the top or bottom will appear stretched. The vertical circuits use feedback extensively to adjust a signal to more or less linear conduction. This maintains picture linearity and corrects picture distortions.

During linearity correction, a sample of the sawtooth signal is fed back to the amplifier. The sample is the opposite polarity of the nonlinear signal; thus, the nonlinearity is canceled out. Feedback also is used to "round off" the peaks in sawtooth signals.

Pincushion Circuit

Some televisions have a vertical pincushion circuit that prevents uneven vertical deflection, as shown in *Figure 7-20*. When the vertical deflection is uneven, the picture appears to pull in toward the center at the top and bottom, and bow out on the sides. The horizontal pincushion effect causes the picture to pull in on the sides.

Note: In older televisions, the feedback capacitor had a very high failure rate.

Figure 7-20. The pincushion effect.

The vertical pincushion circuit controls the current flow to the yoke, which corrects the scan lines at the center, top and bottom of the screen. The correction voltage is taken from the flyback and sent to a transformer that is connected in series to the vertical yoke coils.

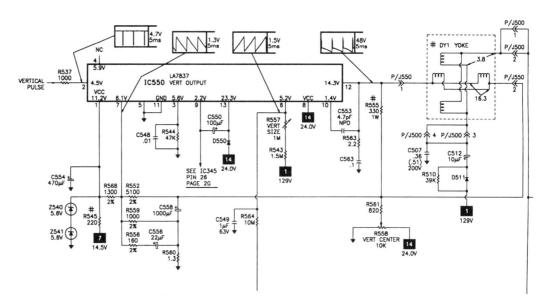

Figure 7-21. A vertical output circuit.

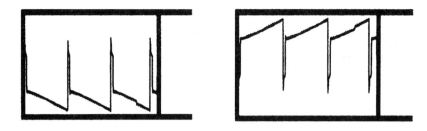

Figure 7-22. Vertical output waveforms.

Vertical Output Circuit

The vertical output circuit, shown in *Figure 7-21*, is a power amplifier that supplies a vertical output waveform to deflect the electron beams to perform the vertical scan and retrace on the CRT's screen. Because of the current in the circuit, the amplifier in most televisions is mounted on a heat sink. The output from the vertical output circuit is a peaked sawtooth, shown in *Figure 7-22*. The vertical output signal is sent to the CRT's vertical deflection yoke coils and is used during scanning. The peaked portion creates the blanking signal that blanks the screen during the vertical retrace.

The output circuit can be directly coupled to the deflection yoke. However, the output circuit can use a resistor, capacitor, transformer or transistor to couple to the deflection yoke.

Figure 7-23. The horizontal bar due to loss of vertical sync.

Troubleshooting the Vertical Deflection Circuits

When you troubleshoot the vertical deflection circuits, use an oscilloscope to look at the waveforms of the signals. Use a DMM to check the input and output voltages.

If the vertical stages are not working properly, the most common symptom is a horizontal white line on the screen and vertical distortion or intermittent vertical scanning. If a horizontal line appears, like the one shown in *Figure 7-23*, and nothing else, turn off the television or turn down the brightness because the CRT can be damaged. If the vertical sweep fails on newer sets, they may blank the raster to prevent CRT damage. In this case the horizontal white line will not appear on the screen unless the master screen control on the horizontal output transformer is adjusted.

With picture problems that are not severe, such as distortion, line pairing or line splitting, or a short picture, use the vertical controls to adjust the picture:
1. Use the vertical height control to adjust the picture height.
2. Use the vertical hold for sync problems. Some television sets have eliminated the need for a vertical hold control.

Figure 7-24. Keystoning.

CHAPTER 7: TROUBLESHOOTING DEFLECTION CIRCUITS

If the control does not adjust the picture properly, look for leaking or open diodes, capacitors, or transistors. A worn potentiometer also can cause a vertical picture problem. In the case of constant line pairing or line splitting, also check the sync separator. If the problem is intermittent, check the high-voltage stages. You could be seeing the results of a voltage discharge or arcing.

If the bottom part of the picture appears to be correct and the top is not, or the other way around, check for faulty push-pull transistors used in the vertical output circuit, or for a faulty capacitor. Disconnect and check the transistors and capacitors.

Distortion also can indicate a linearity problem. A distortion such as keystoning—the picture is smaller on one side than the other, as shown in *Figure 7-24*—can be caused by a defective deflection yoke, as well as incorrect signals from the vertical deflection circuits. However, when the picture spreads at the top or bottom, and compresses at the opposite end, there is probably a linearity problem. *Figure 7-36* shows waveforms that are linear and others that are not.

When the picture rolls, use the vertical hold control to adjust the picture. If the control does not adjust the picture correctly, use an oscilloscope to check the signals into and out of the vertical oscillator. Also, check the sync separator because the vertical oscillator may not oscillate unless it receives sync signals. If the vertical oscillator, the sync separator and all components connected to the two circuits are operating correctly, and all connections are sound, check the vertical drive circuits and yoke.

If the sync separator does not correctly separate the signals, there may be horizontal signals or noise mixed in with the vertical signals. In this case, check the sync separator and sync filter circuits. Then, check the circuits, components and connections in the horizontal sync circuits.

If the vertical deflection IC has been shorted, the supply voltages will probably be lower that usual. If the IC is open, the supply voltages will probably be higher than usual. If the IC output is lower than expected and the supply voltages are the expected values, check the components, especially the external bias components and connections around the IC. If the components and connections are operating properly, the IC is probably bad.

Note: *If the vertical height control does not adjust the height correctly, check all electrolytics in the vertical circuits and capacitors in the feedback circuit.*

Quiz

1. What is the purpose of the peak or spike of the vertical output waveform?
2. What is the proper name of the horizontal oscillator correction circuit?
3. Why does the vertical scan start at the top left for the first field of scan, and the top center for the second field of scan?
4. The horizontal drive signal drives what circuit?
5. What component deflects the electron beams in the CRT?

Key

1. Vertical blanking.
2. APC (automatic phase control) or AFC (automatic frequency control).
3. It allows for interlacing.
4. The horizontal output circuit.
5. The deflection yoke.

Chapter 8
TROUBLESHOOTING HIGH-VOLTAGE CIRCUITS

Note: Make sure you follow the safety guidelines in Chapter 2, **Working Safely**, when working on high-voltage circuits. In particular, read the sections *Avoiding Electrical Shocks When Servicing High-Voltage Circuits and CRTs* and *Avoiding X-Ray Radiation and High-Voltage Limits*. Also, when you troubleshoot high-voltage circuits, make sure you use a high-voltage probe and use an isolation transformer. Do not use a VOM or DMM without high voltage capabilities.

Chapter 8
Troubleshooting High-Voltage Circuits

A television's high-voltage circuits provide the very high voltage needed by the CRT. They also provide additional B+ supplies not normally produced by low-voltage sources. These circuits are much more stable than the power supplies used in earlier televisions because they are more efficient and produce less heat. Also, there are fewer components external to the horizontal output transformer (flyback); therefore, there are fewer places where problems can arise. The high voltage stages are shown in *Figure 8-1*.

There are several types of high-voltage power supplies. This chapter discusses typical high-voltage circuits and provides troubleshooting techniques for these circuits.

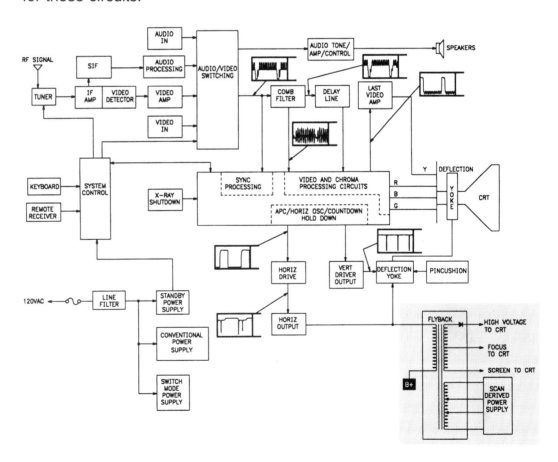

Figure 8-1. The high-voltage stages (shaded).

CHAPTER 8: TROUBLESHOOTING HIGH-VOLTAGE CIRCUITS

When you test the voltages on the high-voltage circuits, compare the values you get with the values on the schematic.

Flyback Transformers

The flyback transformer is an integrated transformer with several low-voltage windings and most often contain molded high-voltage diodes.

The regulated B+ voltage supplied to the horizontal output circuit helps to regulate the output. If the B+ increases or decreases, the output from the high-voltage circuit would increase or decrease, respectively. Additional methods of stabilizing the high voltage are used, including using the flyback to regulate itself, and operating with a certain amount of core saturation to prevent high-voltage buildup.

If the flyback is working, you will find high voltage at the CRT's anode connection. However, before you take the measurement, make sure the high-voltage probe's ground lead is grounded to the chassis CRT dag ground.

To measure the high voltage range, set the brightness and contrast controls to maximum and measure the voltage at the second anode button. Then, set the brightness and contrast controls to minimum and measure the voltage again. This is the range in which the television normally operates.

Troubleshooting a Flyback Transformer

The symptoms of a defective flyback transformer can vary. When you test the flyback transformer, follow these steps:
1. Ground the chassis before troubleshooting the flyback transformer.
2. Unplug the television from the wall outlet.
3. Smell the transformer. A flyback transformer with burned connections or components smells like burned paraffin.
4. If preliminary circuit checks do not show the defect, remove the transformer from the circuit.
5. Use an ohmmeter to check each pin. Refer to the schematic for the expected values.

If any winding has a reading that indicates it is open, replace the transformer.

Shorted Flyback Transformer

When a flyback transformer shorts, the transformer may arc and be very warm. This is because the insulation between the windings can break down, causing high-voltage arcing. Another symptom of a shorted flyback to watch for is excessive brightness on the screen with heavy noise lines due to internal arcing.

Use a variable line transformer to reduce the AC input to 75V. Then, slowly increase the voltage until you detect the arcing. You may be able to see the arcing in a low light setting. Poor solder joints where the flyback is soldered to the board often cause arcing. If a resistance check or the flyback windings indicate any winding of the flyback is open, replace the flyback.

If the flyback is not open, remove the flyback from the chassis and try ringing it using a signal generator. When you ring the flyback, you inject a sawtooth waveform, or ringing pulse, across the primary windings of the flyback and use an oscilloscope to measure the secondary for a waveform, shown in *Figure 8-2*.

Also, you can unsolder the connections and measure the resistance of the transformer's windings. Compare the measured values to those in the appropriate PHOTOFACT.

You can check the focus control by using a high-voltage probe and changing the focus setting. If the focus control is broken, you may have erratic focus voltage and see lines in the picture.

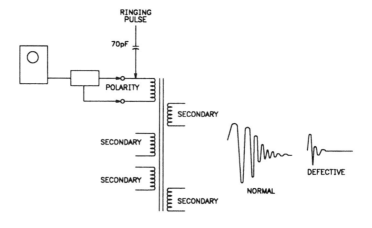

Figure 8-2. Ringing the flyback.

Noisy Transformer

A ticking sound or a high-pitched squeal coming from the television can indicate a problem with the flyback. There might be a loose mounting bolt or the core of the transformer may be vibrating. Locate the faulty transformer and replace it. If the transformer appears to be shut down, reduce the AC input if the high voltage returns, and measure the input and output with a high-voltage probe.

You can hold a rolled-up paper tube up to the transformer to help isolate the sound from the other sounds the chassis is making by using the tube as a stethoscope.

If the transformer is vibrating, it may produce a high-pitched, irritating squeal. Unplug the chassis and inspect the transformer for loose bolts or for a cracked ferrite core. Clean off the old solder from the transformer and from the board. Carefully tighten the bolts, then reposition the transformer and resolder the connections.

Excessive Voltage Level

Excessive voltage from a defective flyback transformer can damage the CRT or chassis circuits. It also can cause the chassis to shut down. A high-voltage shutdown circuit typically monitors a pulse from the flyback. If it exceeds a predetermined level, the horizontal oscillator will stop operating.

An insufficient high-voltage level can cause a focus problem. This condition could be caused by a leaky flyback transformer. However, poor focus also can be caused by faulty CRT connections, or a faulty focus control or low-drive signal. Measure the voltage at the CRT's anode terminal. Then, refer to the television schematic for the correct voltage range. If the voltage is normal, look for the problem in the CRT. If the voltage is low, use signal tracing and test the voltage path back through the high-voltage circuit.

Replacing a Flyback Transformer

If after testing the flyback thoroughly you find that it is deflective, replace it with an exact replacement:
1. Unplug the television from the wall outlet.
2. Discharge the CRT before working on the flyback transformer.
3. If applicable, write down which color lead connects to each terminal and component connected to the transformer.

4. Using a low-wattage soldering iron, slowly heat each connection until the connections are broken and you can lift the transformer. If necessary, clip any leads that are frayed. Note any special routing of the leads.
5. Gently press the new transformer into place.
6. Using your notes, solder the leads and components to the transformer terminals.
7. Visually check all connections for loose solder or sharp solder points that may cause arcing.
8. Using a DMM, check each winding on the transformer. Refer to the schematic for the correct measurements.

Boost Voltage

Boost voltage (200V) is considered to be "high." The boost voltage typically originates in the flyback circuit.

If the boost voltage is not working properly, you might see vertical deflection problems and picture dimming, or other symptoms associated with low-voltage conditions. Look at the schematic and locate the boost. The problem can be a leaky diode, capacitors, or a faulty flyback winding.

X-Ray Protection

X-ray radiation is caused by excessively high voltage. Federal regulations require that television manufacturers protect consumers from X-ray radiation. One of the protection circuits causes the horizontal oscillator circuits to shut down if excessively high voltage is detected. Others may cause the vertical, on/off or power supply circuits to shut down.

A common example of a high-voltage shutdown circuit is shown in *Figure 8-3*. If the circuit detects excessive voltage in the high-voltage circuit, the television shuts down. A sample pulse from the flyback is rectified and fed to one end of a zener diode. If the high voltage increases, so does the sample pulse. If the voltage rises to a predetermined level, the zener typically stops the horizontal circuit from oscillating, shuts down the set, or triggers the horizontal oscillator to reduce the high voltage.

Some televisions have an extremely stable high-voltage circuit or a high-voltage hold-down circuit. In these televisions, shutdown due to excessively high voltage is infrequent.

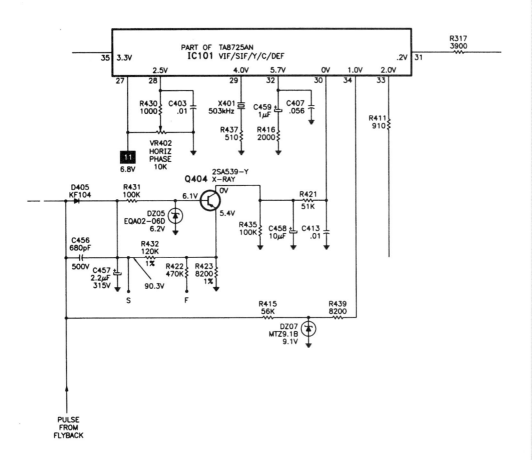

Figure 8-3. A high-voltage shutdown circuit.

If you need to troubleshoot a high-voltage shutdown circuit, use a variable line transformer to reduce the AC input to 75V. Slowly increase the voltage. Measure the sample pulse and output to the high-voltage shutdown circuit. Then, check the components and connections around the circuit. Sometimes it may be necessary to disable the shutdown circuit to locate the problem.

Focus Circuits

High-voltage bleeder resistors can be used to help stabilize high voltage in televisions. The bleeder resistor is most often connected in the low side of the horizontal output transformer high-voltage winding, as shown in *Figure 8-4*, to regulate the first 100 mA, a range that the flyback circuit does not regulate. This helps the flyback operate more efficiently.

The bleeder resistor also keeps the focus voltage in a constant relationship with the CRT's second anode voltage. The focus voltage is approximately 20% of the anode voltage. If the focus voltage is incorrect, the picture will

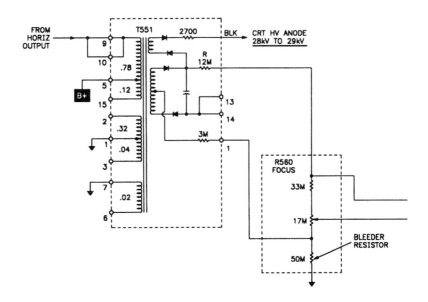

Figure 8-4. A bleeder resistor.

not be correctly focused. If the focus voltage is not present, the raster does not appear, even though you might see a flash when you turn off the television. If the bleeder resistor is an integral part of the flyback transformer, you have to replace the whole flyback transformer. (*Figure 8-5.*)

Bleeder resistors also bleed off the high voltage charge on the CRT when the TV is turned off. This prevents damage to the CRT screen by a small bright dot in the center of the screen, which tends to damage the phosphor on the screen.

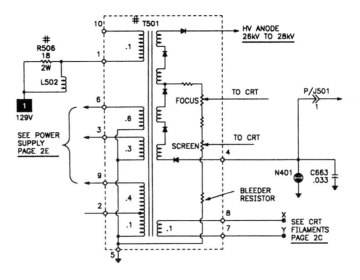

Figure 8-5. A schematic clip showing the current focus bleeder in the flyback.

Automatic Brightness Limiter (ABL) Circuit

Another method of monitoring the high-voltage and keeping the picture in focus while protecting the CRT is the automatic brightness limiter (ABL) circuit, shown in *Figure 8-6*. As its name implies, this circuit limits the current from the flyback to the electron guns in the CRT, thus limiting the picture's brightness and preventing blooming on the screen. The ABL circuit is in the video processing stage and samples the voltage to the CRT's second anode. The circuit uses the sample to determine whether to limit the drive signal to the CRT's electron guns.

Troubleshooting High-Voltage Circuits

Open or shorted components, which results in no raster, are the most common source of problems in the high-voltage circuits. However, with shorted components, you may see the burned components or connections. Also, in the case of a shorted CRT, the high-voltage circuits may not work correctly, or you may see high-voltage arcing in the CRT neck. If components are leaky, causing lower voltage values, the picture may appear dim, out of focus and narrow horizontally.

As you can see, the symptoms of high-voltage problems are similar to problems caused by circuits and components in other stages in the chassis. The main difference is that you must take extra precautions due to the very high voltage potential in order to prevent getting shocked. Therefore, use the following troubleshooting steps to help you determine whether the symptoms you see are caused by a high-voltage circuit or by circuits in another stage.

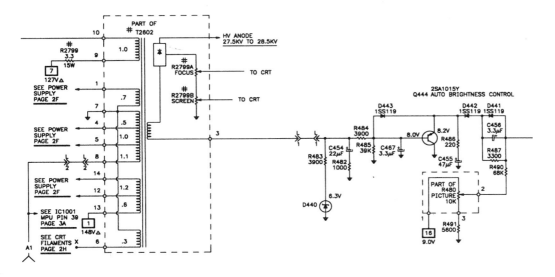

Figure 8-6. A brightness limiter.

In the following steps, use a high-voltage probe or a DMM:
1. Check the output of the horizontal output circuit. If a transistor is leaky, the voltage will be low. If this is the case, the picture will be narrow horizontally and the circuit may show signs of overheating and smell "hot." If a transistor is open, there will be no high voltage. If a transistor is shorted, the fuse will blow or the television will shut down.
2. Check the input to the base of the horizontal output circuit. If there is no input, check the horizontal circuits.

Quiz

1. Why should extra precautions be taken when working around the flyback?
2. What circuit reduces the current of the CRT?
3. What circuit monitors the high voltage level and shuts down the set if the level gets too high?
4. From where does the boost voltage originate?
5. What does the bleeder resistor do when the receiver is turned off?

Key

1. Because of the very high voltage potential.
2. The ABL circuit.
3. The X-ray protection or shutdown circuits.
4. The flyback.
5. It bleeds off the high-voltage charge on the CRT to prevent the CRT from burning a spot in the center of the screen.

Chapter 9
TROUBLESHOOTING TUNER CIRCUITS

Chapter 9
Troubleshooting Tuner Circuits

A television's tuner stage, shown in *Figure 9-1*, selects one channel from all of the signals received by the antenna. Each channel consists of an RF frequency that contains two carriers: the FM sound carrier and the AM video carrier. The tuner circuit adjusts the level of the RF signal and converts it into an IF signal. So, the tuner circuit has one input, the RF frequency, and one output, the IF signal, even though there are two carriers in the RF frequency.

Each of the television stations in a geographic area transmit on a specific frequency. In order to select one signal from all of the possible signals and have all of the other stages in the chassis recognize the selected signal without having to constantly retune, televisions use a tuning method called heterodyning.

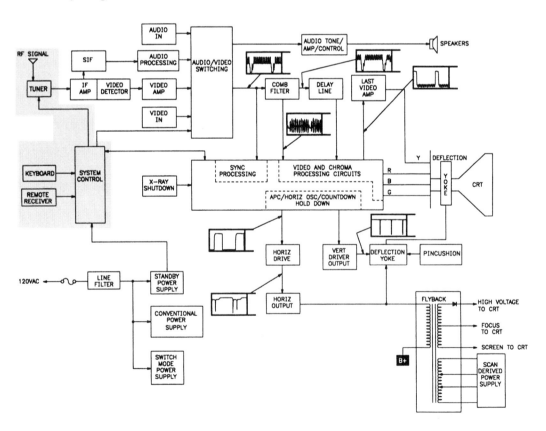

Figure 9-1. The tuner stage (shaded).

CHAPTER 9: TROUBLESHOOTING TUNER CIRCUITS

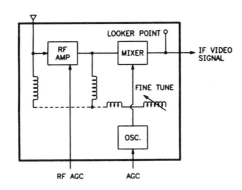

Figure 9-2. A "Generic" tuner circuit.

In heterodyning, when you select a channel, the RF amplifier is tuned to that frequency. The RF amplifier amplifies the signal and outputs it to the mixer stage. Then, the oscillator outputs its signal reference to the mixer. These two input signals are mixed so that the output signal from the mixer is the difference between the two signals, or the intermediate frequency (IF). All of the later stages in the chassis are permanently tuned to the IF so that they do not have to be retuned each time you select a new channel.

There are two types of tuner circuits:
1. A "discrete" tuner is composed of individual components as part of the main board. If a discrete tuner is not working correctly, you can test the components, locate the one that is faulty, and repair it.
2. A modular tuner is a replaceable unit that is easier to replace than try to repair when it is faulty.

Figure 9-2 shows a block diagram for a "generic" tuner circuit.

Manufacturers have developed several types of tuners, including varactor (variable capacitance diode) and frequency synthesis tuners. However, all tuners, including the older mechanical tuners and hybrid tuners, have three main tuner stages: RF amplifier, oscillator and mixer.

RF Amplifier

The RF amplifier:
1. Provides frequency selectivity.
2. Improves the signal-to-noise-ratio.
3. Prevents the radiation from the oscillator from leaking back to the antenna.

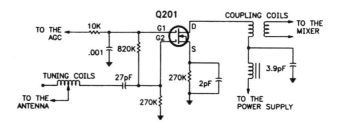

Figure 9-3. An RF amplifier.

The RF amplifier stage, shown in *Figure 9-3*, selects one RF signal from all of the signals being received by the antenna, amplifies it, and outputs it to the mixer. The AGC (automatic gain control) is connected to the RF amplifier and controls the gain of the amplifier by offsetting variations in the carrier signals. The RF amplifier can be a transistor or a FET (field effect transistor).

The television's antenna is part of the RF amplifier circuit. The RF amplifier blocks the signal from the oscillator from leaking back to the antenna. Signal leaking can cause picture interference.

Oscillator

The oscillator stage, like the one shown in *Figure 9-4*, outputs an unmodulated reference signal to the mixer that tracks the desired RF signal at a fixed, offset frequency. The frequency of the reference signal is manufacturer dependent.

Also, tuners have an AFT (automatic fine tuning) circuit connected to the oscillator. The AFT shifts the oscillator frequency as needed to keep the signal in phase with the selected channel frequency. Older televisions have manual fine tuning controls that keep the signal in phase with the channel frequency.

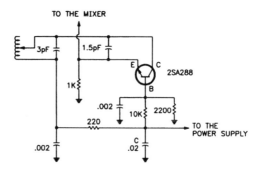

Figure 9-4. The oscillator stage.

Mixer

The mixer stage, like the one in *Figure 9-5*, inputs the RF signal and the oscillator signal, mixes the signals, strips out the extraneous signals from the original RF signal and oscillator signal, and produces the IF signal. The conversion must be performed without distorting the incoming signal. Mixers are sometimes called frequency converters, frequency translators or heterodyne detectors.

The output from the mixer is an IF signal, as shown in *Figure 9-6*. The frequencies represented in the signal are constant so that the signal can be used in later stages in the chassis:
1. 39.75 MHz—The adjacent channel's video carrier.
2. 41.25 MHz—Audio carrier.
3. 41.67 to 42.67 MHz—Color carrier.
4. 45.75 MHz—Video carrier.
5. 47.25 MHz—Adjacent channel's audio carrier.

Tuner Types

Listed below are only three of the types of tuners you will see. The design of most tuners are dependent on the individual manufacturers:
1. A varactor tuner is a diode that acts like a variable capacitor when the diode is biased by a variable voltage supply. The capacitance can be changed over a large range of frequencies by inputting a control voltage from the regulated low-voltage power supply.

    When a channel is selected, a pre-selected voltage for that channel is applied to the diode. This causes the diode to align with the selected frequency. When you fine tune the frequency, the applied voltage is slightly adjusted using a potentiometer on older sets, or by an AFT correction voltage on newer sets, until the diode is completely aligned with the frequency.

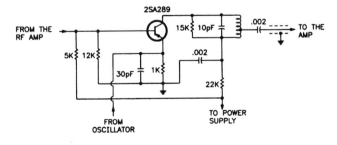

Figure 9-5. The mixer stage.

2. A reactance tuner consists of coils and a capacitor. Each coil is assigned to a tuner frequency. The coils are connected with a multiple contact switch or a series of diodes used as switches. When a new channel is selected, the switch contacts another coil and the channel changes.
3. A frequency synthesis (FS) tuner, also known as digital tuning or quartz tuning, provides up to 127 channels and all of the UHF frequency band. The FS circuit is an IC that contains additional stages such as a phase locked loop (PLL), step generator, a frequency band switch decoder, an AFT, digital sync, and a presence decoder. Refer to the schematic for details about the FS tuner.

Troubleshooting Tuner Circuits

The quickest way to see if the tuner circuits are faulty is to inject an IF signal at the tuner's IF output. If the picture returns to normal, check AFT, B+ supplies, and switching circuits. If they check out okay, replace the tuner.

If the picture does not return to normal, check the IF amplifier, the video processing stages and the power supplies. A defective power supply can cause the channel frequencies to "drift," which results in poor fine tuning and poor reception.

Picture problems can be caused by faulty tuners, as well as other stages in the chassis. The components in the tuners are very critical in value and placement in the circuit. Extreme care must be used when working in the tuner as misalignment can occur, which can result in a very poor or no picture. With any picture problem, if you determine that the tuner is faulty, it is many times less costly to replace the tuner than to try to repair it, plus this will result in a much better picture.

Note: Be careful when working with FET devices because they are easily damaged by static charge. Therefore, when troubleshooting tuner circuits, avoid static charge buildup by using a grounding wrist strap or a static discharge rug.

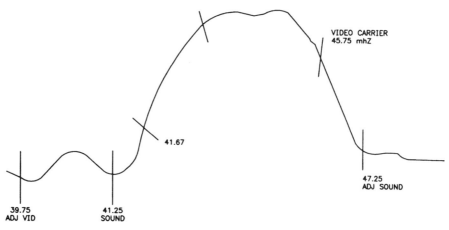

Figure 9-6. An IF signal waveform.

CHAPTER 9: TROUBLESHOOTING TUNER CIRCUITS

Drifting

If the television appears to have drifting frequencies and is not holding one channel, the problem may be in the power supply to the tuner:
1. Disconnect the AFT (automatic fine tuning) circuit.
2. Check the inputs to and outputs from all components (or pins in an IC) that are in the supply voltage path.
3. If the supply voltage varies, the supply voltage circuit may be faulty.
4. If the supply voltage is stable, the tuner may be faulty.

Strong Raster with No Picture or Sound

If you see a strong raster pattern on the screen with no picture, and have faint or no sound, the tuner's mixer might be faulty. This condition also is caused by a faulty IF amplifier or a video processing circuit. If there is no snow, the mixer is probably faulty because it generates most of the raster snow (noise) pattern.

Strong Raster with Weak Picture and Sound

If you see a strong scanning pattern and very weak picture, and hear very faint sound, the RF amplifier or the antenna connection might be faulty. This condition is often accompanied by a hissing sound. If you adjust the contrast, the snow should increase. If you adjust the volume, the hissing sound should increase. This indicates that the mixer and the following stages are operating. Therefore, the RF amplifier is probably faulty.

If the RF amplifier is working correctly, check the AFT, the IF amplifier, and the video circuits.

Weak Raster with Weak Picture and Sound

If you see a weak scanning pattern and very weak picture, the mixer is probably faulty. You also might hear faint sound. This condition also is caused by a faulty IF amplifier.

Weak Raster with No Picture or Sound

If you see a very weak scanning pattern and no picture, or do not hear sound, the oscillator might be faulty. This condition also is caused by a faulty IF amplifier.

No Raster, and No Picture or Sound

If the mixer circuit fails, or if the regulated B+ power supply is faulty, you may not see a scanning pattern (snow), and the sound and picture can range from none to very weak. In this case, inject an IF signal at the mixer's output. If the picture returns to normal, and the B+ power supply is working correctly, the mixer is probably the source of the problem.

Replacing a Modular Tuner

When a modular tuner is faulty—you see problems with color, video or sound, or experience intermittent problems—check all of the soldered connections around the tuner shield. Then, if you need to replace a modular tuner, order an exact replacement for the tuner circuit and follow these steps to remove the old tuner and install the new one:
1. Unplug the chassis from the AC power source.
2. Disconnect the antenna lead and the IF circuit lead.
3. Remove solder from any soldered connections. Make sure you remove any remaining solder from the connections so that the connections are clean for the replacement.
4. Unbolt the circuit from the chassis if it is attached by bolts.
5. Insert the new tuner.
6. Complete any connections according to the instructions that may be included with the replacement.

Quiz

1. How many inputs and outputs does the tuner circuit have if there are two carriers in the RF frequency?
2. Describe the two types of tuner circuits.
3. What is the purpose of the RF amplifier?
4. What is the purpose of the automatic fine tuning (AFT) circuit?
5. Describe the differences between a varactor tuner (a), a reactance tuner (b), and a frequency synthese (FS) tuner (c).

Key

1. One of each: The RF frequency (input) and the IF signal (output).
2. The discrete tuner is made of individual components as part of the main board, and the modular tuner is an individual replaceable unit.
3. It provides frequency selectivity, improves the signal-to-noise ratio, and prevent oscillator radiation from leaking into the antenna.
4. It shifts the oscillator frequency as needed to keep the signal in phase with the selected channel frequency.
5a. A varactor tuner is a diode that acts like a variable capacitor when the diode is biased by a variable voltage supply;
5b. A reactance tuner consists of coils and a capacitor. When a new channel is selected, the switch contacts another coild and the channel changes.
5c. A frequency synthesis (FS) tuner is an IC that contains additional stages. It provides up to 127 channels and all of the UHF frequency band.

Chapter 10
TROUBLESHOOTING SYSTEM CONTROL CIRCUITS

Chapter 10
Troubleshooting System Control Circuits

All of the latest televisions have a system control IC that is connected to the tuner, and controls functions such as maintaining:
1. The television's clock.
2. The front panel controls (also called a keyboard).
3. Channel memory.
4. Fine tuning and picture quality controls.
5. On-screen displays (OSD).
6. Customized settings selected by the owner.

These ICs are simple in some televisions and complex in others. For example, complex systems can contain RF switching, a stereo-decoder mode, and audio and video control circuits.

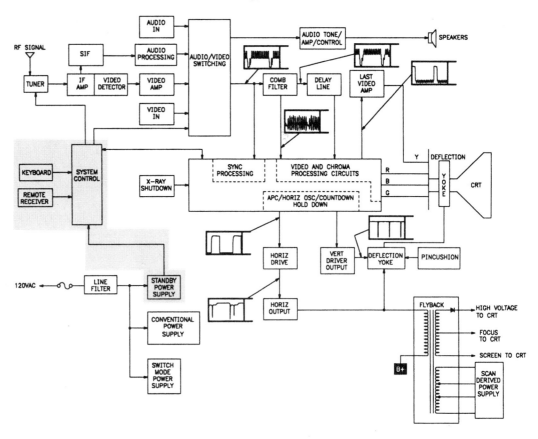

Figure 10-1. Typical system control circuits (shaded).

CHAPTER 10: TROUBLESHOOTING SYSTEM CONTROL CIRCUITS

The system control IC also maintains some of the tuner operations, such as band selection and frequency lock. In many televisions, the system control IC serves as the front-end control for functions processed by circuits later in the chassis, such as the automatic picture-level control and volume control. The precise processes that the system control IC maintains depend on the brand and model of the television.

Also, many times you will see circuits on the IC board that are not connected to other circuits in the chassis. These are features or options that the manufacturer chose not to have available on the specific television model. When you are troubleshooting the system control IC, ignore the circuits that are not used. Refer to the documentation and to the schematic from PHOTO-FACT for information about the specific configuration for the brand and model you are troubleshooting.

The system control circuits can contain a microprocessor(s), diodes, transistors and a frequency synthesizer, which is similar to a variable oscillator. The system control's microprocessor typically has a RAM (random-access memory), and a possible ROM (read-only memory) or a PROM (programmable read-only memory) which store information:
1. RAM stores information that changes, such as the last channel viewed, the last volume level and customized settings.
2. ROM and PROM store the factory settings that do not change, such as the programming for the microprocessor, and on-screen menu and information displays.

If the customer preferences are not stored, the RAM is probably faulty. If the factory settings are not stored, the ROM or PROM might be faulty. In either case, you will probably have to replace the appropriate IC. There are many ways for the system control to control voltage in a television: stepped voltage and high/low voltage are very common. Volume control, in most cases, is an example of stepped voltage because you have a range of values. Switches, like those on the television's keyboard, are examples of high/low voltage control.

Reset Circuit

The reset circuit, shown in *Figure 10-2*, is activated when the AC power is applied and standby voltage is applied to the circuit. The reset pin is an input-only, and is held either high or low. The built-in delay between the power supply and the reset pin allows the system control circuit and its components to be fully powered before the reset is activated and the program starts. The delay time is extremely short.

Note: When you troubleshoot the system control circuits, it is sometimes difficult to locate the problem. Therefore, start your troubleshooting efforts with what is working, not with what is not working. By eliminating the circuits that are working correctly, you can narrow your troubleshooting focus. Also, remember to check the passive components that are connected to the IC, and the solder connections. Sometimes a loose solder connection on a pin, or a leaky or shorted component can cause intermittent and sporadic problems.

Note: Reset is not activated when the reset command from the menu or the remote control is selected. Reset from the menu or remote control only resets the customer preferences. To activate the reset circuit, remove the AC power for 15 seconds, then reapply it.

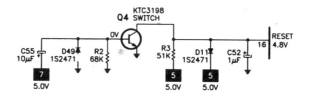

Figure 10-2. A reset circuit.

After the system control circuits and its components are fully powered, reset causes the microprocessor's program to start, which restores the factory settings for the menus and other OSDs, picture control, and so on. The voltage input to the reset circuit is from the standby power supply. The reset circuit can be a transistor or a capacitor.

If the reset signal is missing, incorrect, noisy, or has a slow transition, the microprocessor's program can start incorrectly, causing problems with resetting the factory settings or the microprocessor may not start at all. If the voltage to the reset circuit is incorrect while the AC power is applied, the microprocessor could be resetting continuously, giving a false symptom of a faulty microprocessor IC.

In this case, look for faulty solder connections or cracked traces on the board. Then, test for a faulty open circuit:
1. Before applying AC power, connect an oscilloscope to the reset pin.
2. Apply AC power.
3. Monitor the oscilloscope and look for a change in states. Either the voltage will go from low to high, or it will start high and go low.

If the voltage does not change states, the circuit is faulty.

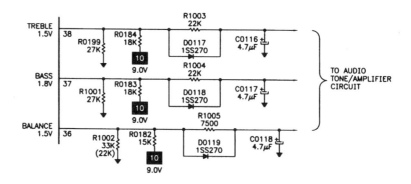

Figure 10-3. An audio mode circuit.

CHAPTER 10: TROUBLESHOOTING SYSTEM CONTROL CIRCUITS

If the voltage changes states, the reset circuit is working properly.

If the circuit appears to be faulty, disconnect the reset pin from the circuit, connect a 1Kohm resistor to the reset pin, and momentarily ground the reset pin. If the settings are not restored, connect the resistor to a 5 volt source. Then, momentarily connect it to the reset pin. If the television's factory settings are not restored or do not operate at all, you will probably have to replace the IC.

Audio Mode Circuits

The volume, balance, bass and treble are controlled in the audio mode circuits, shown in *Figure 10-3*. Audio mode circuits are typically stepped controlled, and the sound usually mutes when the channel is changed. The sound also mutes when the mute function is selected from the remote control or from the keyboard.

The stepped control circuit varies in the number of steps the voltage range is divided into. When the voltage is divided into just a few steps, there is less control over the circuit because the increments are larger. More steps provide greater control and a smoother transition between the levels for volume, bass, treble, and so on. Refer to the PHOTOFACT schematic for the voltage values for the audio mode circuits and the individual features available on the television brand and model you are troubleshooting.

When you troubleshoot the audio mode circuits, your ears are the first tool to use. Listen carefully to the sound. Use the different audio controls on the remote or keyboard. If you hear distortions as you adjust the sound levels (volume, bass, treble, or stereo), try and isolate which audio mode circuits may be faulty.

If you need to inject a signal at any of the audio mode circuits, refer to the PHOTOFACT schematic for the correct voltage levels. Then, inject signals that are a little higher and a little lower than the indicated level to see if the audio changes. For example, if the volume pin indicates 1.5V input on the schematic, test the pin at 1V and 2V to see if the volume changes. A volume control pin is shown in *Figure 10-4*.

Channel Memory (RAM)

Most of the televisions today have an auto channel programming feature. When you turn on the television for the first time, or after it has been unplugged for a while, you may need to issue the auto program command using

Note: The audio mode circuits are not all stepped controlled; they can be controlled by digital ICs transferring data between them. It is much harder to locate a faulty circuit when dealing with audio mode circuits controlled by digital pulses. In these circuits, you can use an oscilloscope to look at the pulse train when the respective circuit is activated.

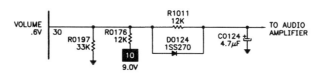

Figure 10-4. A volume control pin.

the remote control or menu, and the tuner then scans for signals that are in an acceptable signal strength range. Then, the tuner transfers the channel numbers to the microprocessor's memory. Some televisions start the auto programming sequence automatically without your intervention. Also, most televisions let you add channels to and delete channels from RAM.

The channel information is stored in the RAM circuit, shown in *Figure 10-5*, and remains there until the television is unplugged or loses power for some length of time. However, a few televisions have small lithium batteries that maintain channel memory, the clock, the last channel viewed, and the last selected volume for a period of time after the television loses power. This feature was designed to hold the television's memory for a short period of time in case of power outages.

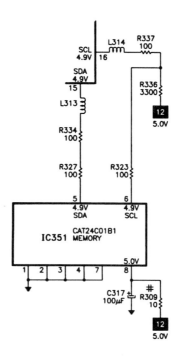

Figure 10-5. A RAM circuit.

When you troubleshoot the RAM circuit, use the auto programming feature to see if the tuner scans the channels and if the RAM circuit "remembers" the channels the tuner selected. If the auto programming commands caused the tuner to scan the channels but the television does not store the channels, the RAM is probably faulty. Try resetting the system to see if the factory settings are restored. Then, attempt to auto program the channels again. If the RAM still does not store the channels, you probably need to replace the memory IC.

Tuner Control Circuits

In newer televisions, channel tuning is available through the tuner control circuits—the microprocessor controls the PLL function used in channel selection. The PLL is a frequency comparison circuit connected by a data line to the tuner. In PLL, the variable frequency oscillator's (VFO) output (frequency and phase) is compared to the fixed frequency reference oscillator's output. If there is a difference in the frequencies, the PLL detects the difference, and increases or decreases the variable oscillator's output until the frequencies and the phases are the same and locked. When the frequency and phase are locked, the channel is clear and undistorted. If the frequency and phase are not locked, the channel will drift—another channel's signal or sound may interfere with the selected channel. You may have to align the IF circuit. Alignment procedures are available upon request from PHOTOFACT if they are not included in the manufacturer's data.

Clock Functions

A television has to have a clock to control timing; otherwise, the system would be chaotic. This clock is different from the 60 Hz clock in the microprocessor that is used for the on-screen time display, sleep timer and alarm clock.

The system control circuit can have one or more clocks. The clock is an oscillator that can be a quartz crystal, a ceramic filter, or a resistor/inductor/capacitor combination.

The clock transmits signals at a fixed repetition pattern over the clock line that connects the clock circuit to the tuner, ROM and other ICs. When there are multiple clocks transmitting signals, each signal is transmitted with a fixed delay or in an set phase relationship with the other clock signals so that the signals do not compete. Unlike the tuner control circuits, instead of being locked in phase, the clock signals are held out of phase at a set number of degrees so that the phase relationships are totally controlled and never compete. Clock circuits are shown in *Figure 10-6*.

Note: *In most cases, the microprocessor unit is connected to the tuner via the data and clock lines. There are very few tests that can be made. To test for normal operation, use an oscilloscope to look for the data transfer between the CPU and the tuner. All PLL circuits are internal to the microprocessor unit in this design.*

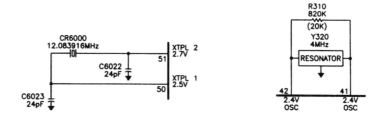

Figure 10-6. Clock circuits.

If the clock circuit is not working, the microprocessor does not work. If the clock circuits are working, but not properly, you can see problems with some of the system control functions, such as the OSD not operating properly.

To test the clock circuit, use an oscilloscope and an impedance probe (also called an X10 probe) to check the oscillator pins for signal activity on the clock line. You also can use a frequency counter to measure the frequency on the clock line. However, be careful not to load the circuit down with the frequency counter. Before measuring the frequency, refer to the appropriate PHOTO-FACT schematic for the television's brand and model, and find the proper frequency for the clock circuit.

If the television is inoperative, and the clock circuits are the source of the problem, there will be no activity on the data lines.

On-Screen Display Circuits

When the on-screen display (OSD) circuits are not working properly, you will see problems with the menus, closed caption information, or other on-screen displays, including the colors and the characters displayed. For example, if menus are partially shown, are scrambled in one area, have parts that appear randomly all over the screen, or do not display at all, check the OSD circuits. If you check the OSD circuit and it is working properly, check the sync sample of the horizontal and vertical circuits going to the microprocessor.

The OSD circuits (*Figure 10-7*) use a character generator to build the numbers and letters that make up the menus and other on-screen displays. The character generator decodes the data signals and transmits them to the CRT, where they are mixed into the video signal in the video processor circuits. The sample horizontal and vertical sync signals going into the IC control the horizontal and vertical location of the on-screen displays. If you see on-screen displays that are in an incorrect position or are partially displayed, use an oscilloscope to check the sync samples.

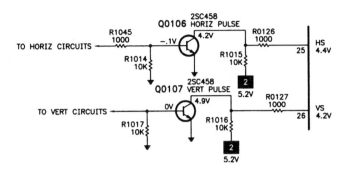

Figure 10-7. On-screen display circuit.

The input to the OSD circuits comes from the ROM. Remember, the ROM stores the factory settings, and the menu, and OSD data. The output from the OSD goes to the video processing circuits, later in the chassis. Therefore, if you see a problem with the OSD, also check the microprocessor circuits and the video processing circuits. If the OSD appears on the screen, but commands from the remote control or keyboard buttons appear not to be working, check the remote's infrared circuits and the keyboard's contacts. If the OSD circuits are the source of the problem, you will probably have to replace the system control IC.

If the television fails to produce a raster scanning pattern, check the blanking circuit in the OSD circuits. An incorrect voltage to the blanking circuit may produce a black raster. Lower voltage to the blanking circuit may produce a lighter raster. If you see only a black raster, check the voltage level to the CRT cathodes and screens.

Closed Caption Circuit

The closed caption circuit controls the closed caption information that is broadcast from the television station along with the video and audio signals. The closed caption circuit may be a separate circuit or integrated in the system control IC.

The closed caption information is encoded on the vertical blanking bar. When there is closed caption available, the closed caption decoder circuit decodes the signals and converts them to red, blue and green signals to be added to the RGB signals being sent to the CRT. The converted signals are output to the microprocessor. Then, the decoded closed caption signal is output to the video processing circuits via the OSD circuits.

Like the OSD circuit, the closed caption circuits use a character generator to build the numbers and letters that make up the menus and other on-screen displays. The character generator decodes the data signals and transmits them to the CRT, where they are scanned to the screen just as the television picture is scanned to the screen. The sample horizontal and vertical sync circuits in the IC control the horizontal and vertical location of the closed caption.

If the closed caption information does not appear on the screen, first check to make sure that the program being received has closed caption information. If closed caption information is available but is not appearing, appearing in the wrong location on the screen, or is distorted, use an oscilloscope to check the sync samples. Also, check the OSD circuits.

Picture Control Circuits

The picture control circuits control the appearance (tint, color, contrast, brightness and sharpness) of the picture displayed on the screen. In most televisions, the picture control circuits are connected to the video processing circuits.

If the picture appears to have the incorrect tint, color, contrast, brightness or sharpness, and you want to check the picture control circuits, apply voltage starting with the voltage level indicated on the schematic. Then, step-up or step-down the voltage to the pin and look for a change in the picture on the screen, If you do not see a difference, the picture control circuits may be faulty; however, these circuits are rarely faulty.

Remote Control and Keyboard Circuits

The remote control unit transmits infrared signals to the infrared receiver on the front of the television. The infrared receiver circuits (usually a preamplifier) amplify the signals and convert them to signals the system control circuits can decode (*Figure 10-8*). These signals are output to the "remote-in" pin, to the microprocessor.

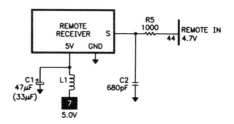

Figure 10-8. Infrared receiver circuits.

CHAPTER 10: TROUBLESHOOTING SYSTEM CONTROL CIRCUITS

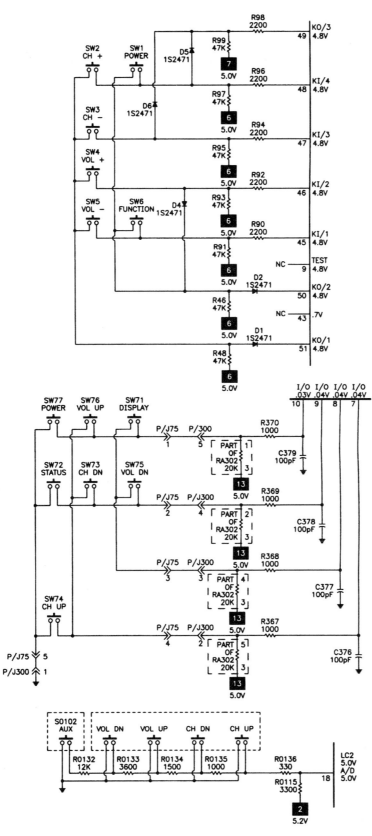

Figure 10-9. Various types of keyboard circuits.

If the remote control does not work, but the keyboard (front panel control) does, first check the batteries in the remote control. If the batteries are fresh, use an infrared card to test the remote. This card can tell you whether the problem is in the remote control or the receiver. If the remote is determined to be good, you can use an oscilloscope to measure the signal at the preamplifier. If there is no signal, the preamplifier may be faulty. If one function (command) in the remote does not work correctly, a good cleaning of the remote control key pad may fix the problem.

There are many types of keyboard circuits; a few are shown in *Figure 10-9*. The precise way a keyboard circuit is interpreted by the microprocessor is manufacturer dependent. They all seem to have a few things in common:
1. The switches are momentary contacts.
2. They normally have a source portage, though it may differ from circuit from circuit.

A keyboard circuit is usually easy to test. Testing can be done either at the switch or at the appropriate pins on the microprocessor. It is sometimes necessary to test both places.

Power Supplies

The microprocessor draws most of its power from the standby power supply. The power is supplied at various points throughout the circuit. Refer to the PHOTOFACT schematic for the specific voltage requirements of each circuit in the IC and for the specific power supplies. Also, refer to Chapter 4 and Chapter 8 in this book for power supply problems.

When you check the power supplies, follow these guidelines:
1. Remove the AC power to reset the microprocessor.
2. Apply the AC power.
3. Check standby power supply to the system control IC.
4. Check the voltage at the power pin for a change of state when the power button is pressed.

If you turn on the television, and it does not come on, check the power supply, the on/off relay circuit, the horizontal circuits, and the blanking circuit.

General Troubleshooting Techniques

Note: As you troubleshoot the system control circuits, refer to the PHOTOFACT schematic for the television model and brand you are working on for the values specific to that television.

If you suspect that the system control circuits are defective, there are quick and inexpensive methods of troubleshooting the IC. First, determine what test can be performed to determine which circuits are good. For example, if an A/V switching circuit is not switching, disconnect the analog line from the microprocessor unit and inject an appropriate voltage to the line for testing. If the A/V switching circuit will switch, then you know that the microprocessor unit is not sending out the proper voltages for switching. Always check any onscreen service menus included in PHOTOFACT, and use them to determine if the microprocessor unit needs to be set up again.

Quiz

1. Describe the difference between RAM, ROM and PROM.
2. What is the purpose of audio mode circuits?
3. What do all keyboard circuits have in common?

Key

1. RAM (random-access memory) stores information that changes; ROM (read-only memory) and PROM (programmable read-only memory) store the factory settings that do not change.
2. Audio mode circuits control volume, balance, bass and treble.
3. The switches are momentary contacts, and they normally have a source portage.

Appendix A
TROUBLESHOOTING SYMPTOMS

Appendix A
Troubleshooting Symptoms

This appendix describes symptoms of problems you might encounter. Then, you are told which chapters contain the troubleshooting and repair guidelines for that symptom.

General Television Problems

Arcing. (See Chapter 2, Chapter 4 and Chapter 8.)
Check:
1. The high voltage circuits.
2. The anodes in the CRT.
3. The flyback transformer.
4. For excessive voltage, especially at the CRT anodes.
5. For dirty connections.
6. The CRT for cracks.
7. The high-voltage cable and plug.

B+ fuse blows. (See Chapter 7.)
Check:
1. For a shorted transistor or diode.
2. High voltage circuits.
3. The power supply, especially for shorted diodes or filter capacitors, open diodes, filter capacitors, dropping resistors, fuses or switches, leaky filter capacitors or a shorted transformer or choke.
4. For overloaded voltage sources.

Chassis shutdown. (See Chapter 8.)
Check the flyback transformer and high-voltage circuits (*Figure A-1*).

Frequency (channel) drift. (See Chapter 9.)
Check the:
1. IF amplifier.
2. Tuner (*Figure A-2*).
3. Power supply to the tuner.

Front panel controls (keyboard) do not work. (See Chapter 10.)
Check the keyboard contacts and power supply to the system control circuits.

APPENDIX A: TROUBLESHOOTING SYMPTOMS

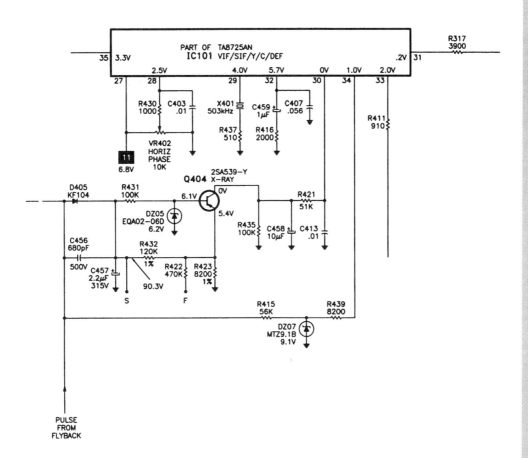

Figure A-1. A high-voltage shutdown circuit.

Interference in video. (See Chapter 5 and Chapter 10.)
Check the:
1. Video IF circuits.
2. The video amplifier (*Figure A-3*).
3. Tuner alignment.

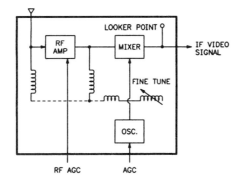

Figure A-2. A tuner circuit.

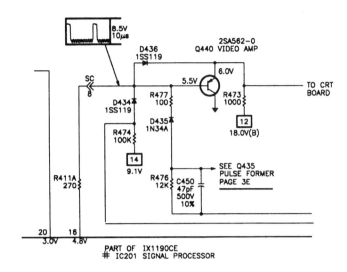

Figure A-3. The last video amplifier.

No channel memory; fails to lock in on one channel. (See Chapter 10.)
Check the:
1. AFT circuits.
2. RAM circuit in the system control IC (*Figure A-4*).
3. Tuner alignment.

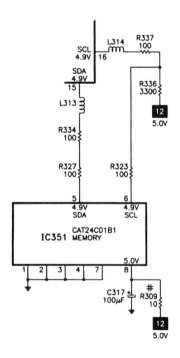

Figure A-4. A RAM circuit.

APPENDIX A: TROUBLESHOOTING SYMPTOMS

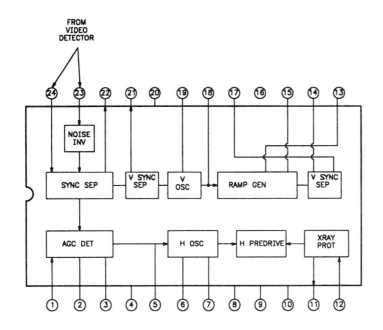

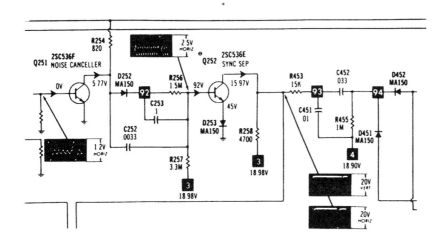

Figure A-5. A sync separator circuit.

Noise in the sync signal. (See Chapter 7.)
Check the:
1. Sync separator (*Figure A-5*).
2. Video detection circuits.
3. Noise reducer (canceller) circuit.

On-screen display (OSD) problems. (See Chapter 10.)
Check the system control circuits, especially the OSD circuits.

Power supply turns off. (See Chapter 4.)
Check:
1. The capacitors, resistors and inductors in the power supply.
2. For excessive voltage.
3. For loss of filtering due to faulty capacitors.
4. For loss of output voltage because of a shorted output capacitor, open-circuited resistor or an open choke.

Raster, picture and sound problems. (Chapter 4, Chapter 5, Chapter 8 and Chapter 9.)
Check:
1. The tuner circuits.
2. The IF amplifier (*Figure A-6*).
3. The AGC.
4. The video detector.
5. The tuner's mixer.
6. The video processing.
7. For a short in the video amplifier circuits.
8. The power supplies.
9. For very low or very high voltage to the CRT.
10. For shorted or overheated components in the high voltage circuits.

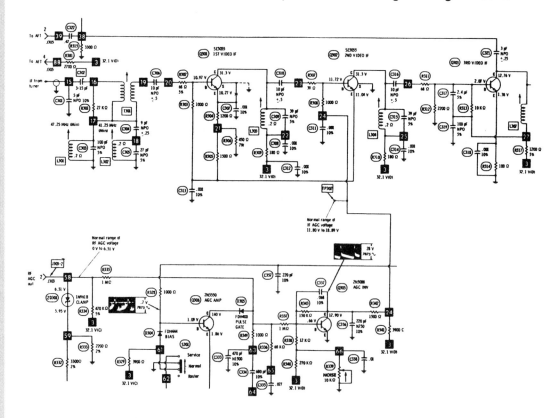

Figure A-6. An IF amplifier connected to an AGC.

APPENDIX A: TROUBLESHOOTING SYMPTOMS

Receiving no channels. (See Chapter 9, Chapter 10.)
Check the:
1. Tuner circuits.
2. Antenna connection.
3. Solder connections in the tuner and system control IC.

Remote control does not work. (See Chapter 10.)
Check the:
1. Batteries in the remote control.
2. Infrared receiver circuits in the system control IC.

Reset does not work. (See Chapter 10.)
Check:
1. The reset circuit in the system control IC (*Figure A-7*).
2. For excessively low voltage.
3. For faulty solder connections or cracked traces in the system control circuit.

Television is inoperative. (See Chapter 4, Chapter 7, Chapter 8 and Chapter 10.)
Check the:
1. Power line cord and plug.
2. Television's fuses.
3. System control circuits, especially the clock and power pin.
4. Standby power supplies.
5. High-voltage protection circuit.

Television turns on, then turns off. (See Chapter 4.)
Check for a voltage overload or a high-voltage shutdown.

Ticking sound or a high-pitched squeal. (See Chapter 8.)
Check the flyback.

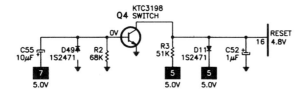

Figure A-7. A reset circuit.

Picture Symptoms

A spot (blotch) appears on the screen. (See Chapter 5.)
Check the CRT's shadow mask or the degaussing coil.

All colors are incorrect. (See Chapter 5.)
Check the:
1. Tint control.
2. Chroma processing circuits.
3. CRT.
4. Kine board.

Black bars drift from bottom to top. (See Chapter 4.)
Check the filters in the power supply (*Figure A-8*).

Blanking bar appears on the screen. (See Chapter 7.)
Check the sync separator circuit (*Figure A-9*).

Blooming. (See Chapter 5 and Chapter 7.)
Check for excessively high voltage, boost, or screen voltage.

"Breathing" picture, vertically and horizontally unstable. (See Chapter 4.)
Check the power supply regulation.

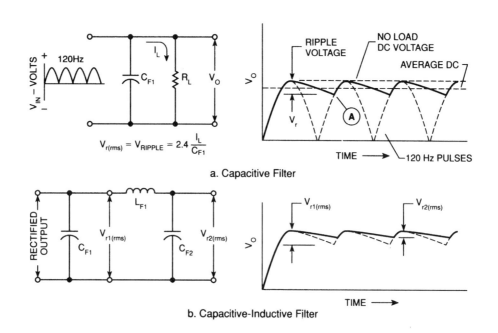

Figure A-8. Filtering.
(Reprinted by permission from *Power Supplies: Projects for the Hobbyist and Technician*
©1991, 1992, Master Publishing, Inc.)

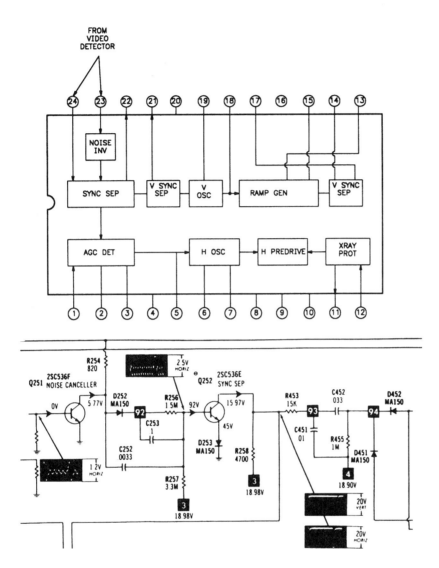

Figure A-9. A sync separator, internal to IC.

Brightness problems.
Check:
1. For a high voltage problem.
2. The luminance circuit.
3. The output from the final video amplifier circuit.

Color bleeding. (See Chapter 5.)
Check the comb filter and screen voltage.

Color is too intense. (See Chapter 5.)
Check:
1. The color control.
2. The color killer.
3. The automatic chroma control (ACC).

Colors appear dim. (See Chapter 5.)
Check the:
1. Electron guns for oxidation.
2. CRT for shorting.
3. Luminance circuit.
4. Output from the chroma amplifier circuit.

Horizontal and vertical pincushioning. (See Chapter 7.)
Check the pincushion circuit.

Horizontal white line on the screen. (See Chapter 7.)
Check the:
1. Vertical circuits (*Figure A-10*).
2. Deflection yoke.

Keystoning. (See Chapter 7.)
Check the:
1. Deflection yoke.
2. Vertical deflection circuits.
3. Linearity.

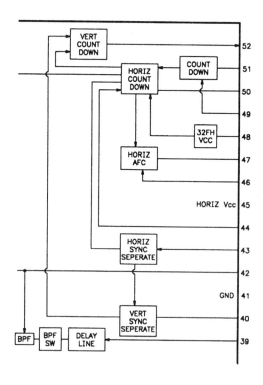

Figure A-10. *A vertical countdown circuit.*

APPENDIX A: TROUBLESHOOTING SYMPTOMS

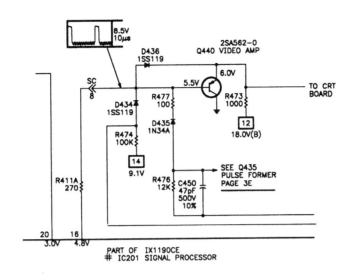

Figure A-11. The last video amplifier.

Line pairing or line splitting. (See Chapter 7.)
Check:
1. For leaking or open diodes, capacitors, or transistors.
2. For a worn potentiometer.
3. The video processing circuits for open connections.
4. The sync separator.
5. The high-voltage stages.

Little control of brightness, focus or contrast. (See Chapter 5, Chapter 8 and Chapter 10.)
Check:
1. For low voltage at the heater element.
2. For excessively high voltage.
3. The luminance circuit.
4. Output from the final video amplifier circuit (*Figure A-11*).

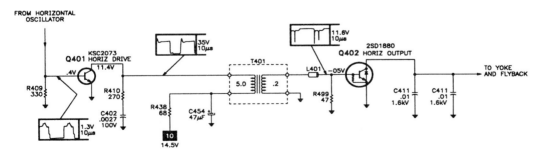

Figure A-12. The horizontal deflection (drive) circuit.

Loss of horizontal and vertical sync. (See Chapter 7.)
Check the:
1. Sync separator.
2. Vertical deflection circuits.
3. Horizontal deflection circuits (*Figure A-12*).

Loss of horizontal sync; picture tears horizontally. (See Chapter 5 and Chapter 7.)
Check the:
1. APC (automatic phase control) circuit.
2. Sync separator circuit.
3. Horizontal oscillator (*Figure A-13*).

Loss of vertical sync; picture rolls up or rolls down. (See Chapter 5 and Chapter 7.)
Check the:
1. Vertical deflection circuits.
2. Sync separator.

Misconvergence problems. (See Chapter 5.)
Adjust the convergence.

Negative picture. (See Chapter 5.)
Check the polarity of the diode(s) in the video detector.

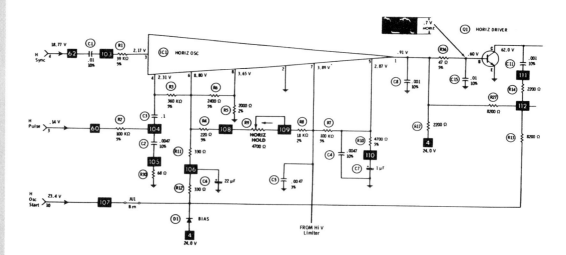

Figure A-13. A horizontal oscillator circuit.

APPENDIX A: TROUBLESHOOTING SYMPTOMS

No color. (See Chapter 5.)
Check the:
1. Color 3.58 MHz oscillator.
2. Comb filter.
3. Color sync amplifier.
4. Color killer.

No picture and no sound. (See Chapter 5 and Chapter 7.)
Check:
1. The video IF amplifier.
2. For an open or leaky capacitor, transistor or damper diode in the horizontal output circuit.

One color is incorrect. (See Chapter 5.)
Check the:
1. Tint control.
2. Color signal demodulator.

One or more colors are lost. (See Chapter 5.)
Check the:
1. Color guns in the CRT.
2. Color controls.
3. Color amplifier.
4. Color signal demodulator.
5. Transistor for the missing color on the Kine board.

Picture does not fill the screen. (See Chapter 4.)
Check the vertical deflection circuits and for voltage that is too low.

Picture folds or weaves, or has width problems. (See Chapter 7.)
Check for:
1. A drop in the voltage pulses to the deflection yoke coil.
2. A short in a deflection coil.

Picture is not linear, the top or bottom appear stretched. (See Chapter 7.)
Check the:
1. Vertical oscillator.
2. Vertical output circuit (*Figure A-14*).
3. Vertical deflection circuits.

Picture looks like a cartoon; no luminance. (See Chapter 5.)
Check the video processing circuits or the CRT circuits.

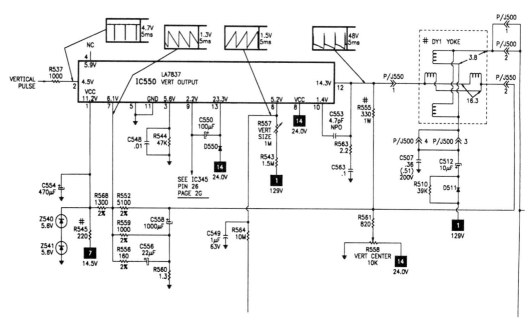

Figure A-14. A vertical output circuit.

Picture narrows horizontally. (See Chapter 8.)
Check for low B+ voltage.

PIP (picture-in-picture) problems. (See Chapter 5.)
Check the:
1. PIP circuit.
2. A/V switching circuits (*Figure A-15*).
3. Supply voltages to the PIP circuit (*Figure A-16*).
4. Video, luminance and color signals at the PIP's input stage.
5. Sync signal processor.
6. Horizontal and vertical deflection circuits.

Reduced focus or intermittent focus problems. (See Chapter 5, Chapter 7 and Chapter 8.)
Check:
1. The flyback circuit.
2. For an overloaded circuit.
3. For a defective picture tube socket or improper focus voltage.
4. The focus control.
5. For a leaky spark gap assembly.
6. The CRT.

Vertical scan, no horizontal scan; white vertical line appears on the screen. (See Chapter 7.)
Check the horizontal output circuit.

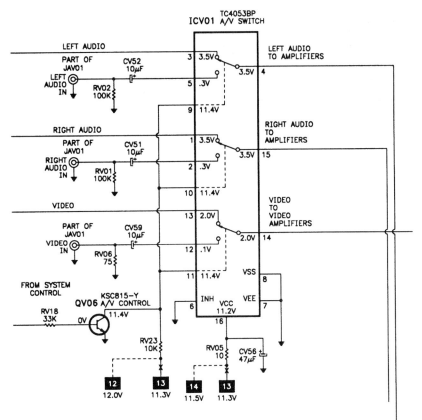

Figure A-15. An A/V switching circuit.

Vertical sync signals in the blanking bar. (See Chapter 7.) Check the sync separator.

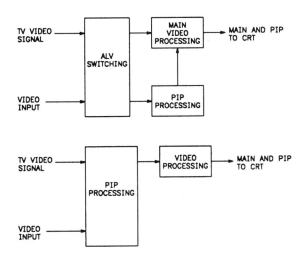

Figure A-16. A PIP block diagram.

Weak or dim picture, reduced brightness. (See Chapter 7 and Chapter 8.)
Check:
1. For low voltage at the heater element.
2. For a voltage source that is too low.

Audio Symptoms

A weak hissing sound that can be increased or decreased using the volume control. (See Chapter 5.)
Check the SIF amplifier.

Crackling or popping. (See Chapter 6.)
Check :
1. The filter capacitors.
2. The power supplies.
3. For faulty resistors.

Distorted sound or intermittent sound. (See Chapter 6.)
Check:
1. For a faulty transistor or resistor.
2. The SIF amplifier.
3. The audio processing IC (*Figure A-17*).
4. The detector coil.
5. For loose wiring and cold solders.

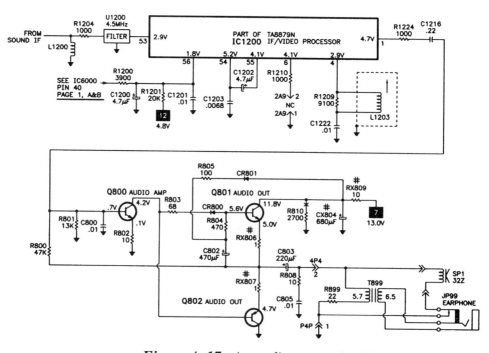

Figure A-17. An audio processing IC.

APPENDIX A: TROUBLESHOOTING SYMPTOMS

Hum, putt-putt, or buzz sounds accompanied by a running sync. (See Chapter 4 and Chapter 6.)
Check the:
1. Sync separator.
2. AGC circuit.
3. Low-voltage power supply circuits.
4. Filter capacitor.
5. B+ power supply.

Interference in the audio signal. (See Chapter 6 and Chapter 10.)
Check the:
1. Audio circuits.
2. Video detector.
3. Tuner circuits.

No second audio program (SAP). (See Chapter 6.)
Check the:
1. Demodulation in the SAP circuit.
2. Tuner circuits.

No sound, no picture, but there is a raster, and a hum or background noise. (See Chapter 5.)
Check the:
1. AGC circuit.
2. Tuner circuits.
3. Video detection circuit (*Figure A-18*).
4. IF circuit.

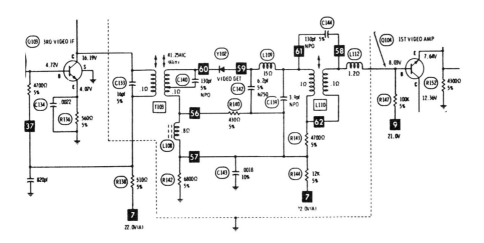

Figure A-18. A diode used as a video detector.

No sound at all. (See Chapter 6.)
Check the:
1. Speakers.
2. Volume control circuits, including the audio mode circuits in the system control.
3. Audio processing circuits.
4. IF amplifier.
5. Video detection circuit.
6. Mute circuit.
7. Audio amplifier.

No stereo. (See Chapter 6.)
Check the:
1. Stereo processing IC.
2. L-R demodulator circuit.

Poor sound quality accompanied by an unstable picture. (See Chapter 6.)
Check the:
1. Filter regulator, capacitors, transistors, resistors, sound alignment in the audio processing circuits.
2. AGC circuits.
3. Tuner alignment.

Ringing sound associated with a picture problem. (See Chapter 7.)
Check for an overloaded circuit.

Sound and picture do not track properly. (See Chapter 6.)
Check the detector coil or sound alignment.

Sound, but no picture or raster. (See Chapter 5.)
Check the:
1. Horizontal circuits and hot transistor.
2. Damper diode.
3. Video processing circuits.
4. CRT.

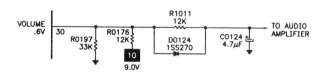

Figure A-19. A volume control pin.

APPENDIX A: TROUBLESHOOTING SYMPTOMS

Squealing or whistling, distorted or intermittent sound, or no sound, even at maximum volume. (See Chapter 6.)
Check:
1. The filter regulator, capacitors, transistors, resistors, sound alignment in the audio processing circuits.
2. The speaker's voice coil or the speaker cone.
3. The detector coil.
4. For loose wiring.
5. All solder connections.
6. The power supplies.

Volume control problems. (See Chapter 10.)
Check the:
1. Audio mode circuits.
2. Power supplies.
3. Volume pin (*Figure A-19*).

Weak sound. (See Chapter 6.)
Check:
1. For a leaky or open transistor or capacitor.
2. For a resistance change or voltage that is too high or too low.
3. The speaker cones.
4. The detector coil.

Appendix B
PHOTOFACT

Appendix B
PHOTOFACT

Throughout this book there are references to PHOTOFACT, Howard W. Sams' technical repair data for TVs. This appendix provides an outline for PHOTOFACT: how they are made, how they should be used, and how to order one.

The main advantage that PHOTOFACT offers the service technician is schematic layout. Manufacturer schematics are laid out by board, while PHOTOFACT is laid out by signal with the power supply separated and drawn as a single circuit. The advantage to the technician is that any problem area is easy to find and troubleshoot on the schematic. The audio portion is always on the top baseline of page 2 with a note describing where to find any additional audio information if space does not permit the completion of that section (stereo, inputs and switching, etc.).

The video signal is on the second baseline; the sync and vertical signals are on the third baseline; and the horizontal signal is on the bottom baseline. Additional circuitry (microprocessor, tuner, remote transmitter, etc.) is covered on separate pages with notes on each end of any connection clearly showing how it is tied back to the main schematic. Since every PHOTOFACT is laid out in this manner, a service technician saves time by not having to search all over the schematic if he has a vertical problem.

The second advantage of PHOTOFACT is accuracy. Every set is wire checked to determine how each component is connected. The technician is supplied with a placement chart and illustrations to aid in locating components, voltage points, adjustments, etc.

Another advantage is a more complete parts list, with not only manufacturers' part numbers but type numbers, replacements, and alternates when appropriate.

Another advantage of PHOTOFACT is that they cover most TVs on the market and service information is available from your local distributor. You don't have to contact the manufacturer of that particular set and wait for service data.

PHOTOFACT schematics are in a consistent layout no matter which brand of TV you are working on. Manufacturers' schematics are very different in content and layout for each brand. Every time you change brand informa-

APPENDIX B: PHOTOFACT

tion, it takes time to find the are of the circuit you want to check on the manufacturer schematic.

PHOTOFACT power supply schematics are complete, drawn as a separate schematic showing where the voltages originate for the entire set. Manufacturers schematics for the power supply may be located in one area or scattered anywhere on the schematic. Sometimes they are difficult to locate.

PHOTOFACT information is consistent in content regardless of the brand name. We supply all the information a technician needs to repair a set. Some manufacturers supply technical training manual-type information requiring you to look through many pages of information to find the problem. Some manufacturers only give voltages at the power supply points, so you must guess the voltage on the transistors and IC pins. Some manufacturers show wave forms at key points and others do not.

PHOTOFACT has a customized parts list with recommended replacements for semiconductors and resistors. We show only the items that are needed for safety reasons or that are special items that a well-stocked service shop wouldn't have in their normal inventory. With manufacturers information, you often waste time looking through several pages of components that are common value items that most service shops do not order from the manufacturer. These items can be ordered from a local distributor.

PHOTOFACT schematics are generated from a production line set that is in stores for sale. The schematic is wire checked point to point to ensure accuracy. Manufacturers' schematics are often produced from engineering drawings before the sets are actually produced. If there are any production line changes in a circuit, they often do not show up on the manufacturers schematic. Several months after the set is produced, manufacturers often issue a correction bulletin that must be filed with the original information. If these bulletins are misplaced, you may be unable to correct the faulty circuits. Manufacturers shadowgrams are often inaccurate and misleading.

Using a television receiver as an example, here are a few of the steps that go into the production of a PHOTOFACT.

The receiver is checked to make sure there are no defects. Thermocouples are then connected to certain components and various points on the chassis, to measure the temperatures encountered during four hours of normal operation.

The cabinet back is removed, and the complete set—with chassis still in the cabinet—is taken to the photo lab for front- and rear-view photos.

The receiver is returned to the disassembly stations where, prior to removal of the chassis, it is again operated as a unit to obtain the data needed to determine which of the available replacement picture tubes are suitable for use with the receiver. Such data includes the operating voltages and the physical dimensions of the unit, as well as any special mounting requirements. At the same time, the physical and electrical characteristics of the speaker(s) and antennas are noted and compared with those of available replacements in order to establish which will meet or exceed replacement specifications.

Prior to removing the chassis, the convergent board and convergence yoke assemblies are analyzed to establish the most effective and accurate setup procedures, including gray-scale purity and convergence adjustments.

The chassis is then removed from the cabinet for easier access to key points in the circuitry and reconnected to the picture tube through extension cables. The chassis is turned on and, with no external signal applied, is adjusted to produce a normal raster. (This establishes valid operating conditions which can be easily duplicated at any service shop.) DC voltages on the elements of the tube, transistors, and at other key points are measured and recorded for later application to the PHOTOFACT schematic.

A test signal is applied to the receiver and the range of AGC voltages, and any other control voltages which vary according to the strength of the receiver signal, are measured and recorded. Color signals from one of a number of available keyed color-bar generators are used for color analysis.

Resistance between the transistor elements and ground are measured and recorded in the next step. Resistance measurements in solid-state circuitry are obtained with an ohmmeter that has a maximum of .08 volts between the probe tips. Such measurement criteria are noted in PHOTOFACT.

The flyback, the yoke, and the vertical-output transformer are then analyzed. Inductance and DC resistance are measured and recorded, and all electrical and physical characteristics are compared with those of available component designs to determine the most suitable replacements.

After the replacements for each part have been selected, they are installed in the chassis and the chassis is operated to ensure that the selected re-

APPENDIX B: PHOTOFACT

placements actually perform satisfactorily. Pulses, voltages, and currents are measured and compared, and a visual check of the raster is made.

Next, the set is aligned, if needed, using post injection on the markers, which is the marker-application method designed into most sweep/marker generators. The test of alignment instruments used at the alignment and other analysis stations are periodically rotated among the major makes of test instruments available to service technicians. This ensures that the alignment and adjustment instruction and all other data published in PHOTOFACT are compatible with the test instruments available to service shops.

Using available test equipment commonly found on a typical technician workbench, the alignment specialists record the detailed information needed to later prepare step-by-step instructions for performing video IF and chroma circuitry alignment.

After determining that all tuned circuits are properly aligned, the alignment specialists digitally capture the response curves, including appropriate markers. These images of the actual response curves produced by the chassis are included in the chassis alignment section of the PHOTOFACT folder.

The alignment specialist then photographs the waveforms produced at key test points in the chassis. He also measures and records the peak-to-peak voltage level of each waveform; these are later stripped onto the waveform photos before they are placed on the PHOTOFACT schematic.

At this point, all dynamic measurements of the chassis have been completed and the chassis is then further disassembled.

The chassis, tuners, and associated subassemblies are returned to the disassembly area for preparation and photography. This preparation includes removal of all shields so that any circuitry and components that normally are hidden beneath them are visible in the photographs and can be labeled for identification. The disassembled chassis and electrical subassemblies of the receiver are photographed in as many positions as required to properly display and identify all circuitry and components.

Using the manufacturer data as a guide, a technician maps out each circuit and notes the actual electrical location and value of each component. He then adapts the image to the uniform signal-flow layout used in PHOTOFACT schematics. When the circuitry or a component on the manufacturer schematic differs from that actually in the chassis, the information on the manufacturer's schematic is indicated on the PHOTOFACT schematic as an alternate circuit or component.

At this point in the process, images of the chassis, circuit boards, and other subassemblies have been developed and produced. These images have component numbers recorded on them to assure that the callouts on the image will correspond to those on the PHOTOFACT schematic. Each circuit and component is checked at least three times. The repetition—which is part of the PHOTOFACT technique—helps ensure accuracy. A small duplicate of the schematic is also produced, which is returned to the PHOTOFACT analysis line and placed with the chassis and associated subassemblies so that future reference to circuitry and components during subsequent analysis will correspond to the data on the finalized PHOTOFACT schematic.

The chassis and all other subassemblies of the receiver are then processed through the individual component analysis section. Here, the characteristics of all components are analyzed to determine which of the available replacement-parts manufacturers equal or exceed the ratings of the originals. The most suitable of the available replacement components are listed as replacements in PHOTOFACT.

Finally, all of the data which has been gathered during the PHOTOFACT analysis process is compiled, and all information and illustrations are rechecked for accuracy and completeness.

At this stage in the PHOTOFACT process, all information, schematics, and illustrations have been complete, verified, rechecked, and are ready to be assembled and printed into a PHOTOFACT folder. When the PHOTOFACT folders have been printed, they are delivered to the shipping/distribution department where they are expedited to the field.

Selecting which TVs to cover in PHOTOFACT is not done by chance. It is a full-time job requiring an in-depth knowledge of both the manufacturers and their markets. In fact, Howard W. Sams has one staff member whose only job is to gather data and create a database to help Sams' technicians decide which TVs to cover.

Because it is almost impossible to reliably forecast breakdown rates, market reports, historical information such as previous repair track records, and customer phone requests are used to get as complete a picture of the universe of TVs as possible. Even with all this data, however, the most important issue in determining which TVs to cover is whether or not the service

APPENDIX B: PHOTOFACT

technician will be servicing a particular TV. Howard W. Sams' technicians then apply their 50 years of experience in making coverage selections, and give complete coverage to the most popular brand names.

The chassis is the heart of any TV, and it is the job of the service technician to keep that heart beating. But before major surgery can be performed, it is necessary to have the right information to ensure that the operation will be a success. First, it is important to go to the source to find out which PHOTOFACT set is needed, either through an electronics distributor that carries PHOTOFACT or by ordering one directly from Howard W. Sams. It is vital to be certain that the correct PHOTOFACT set is available. Typically, ordering PHOTOFACT sets by chassis number alone is a major cause of errors.

Everyone who works on TVs knows that the same basic chassis can be used in a number of different models. However, just because two models have the same chassis number does not mean that the chassis are identical in both models. For example:
1. The tuner section can be different.
2. Some component values can differ.
3. One model may have a stereo board while another model is mono.

In some cases, the same chassis is used in several different brand name TVs. Each different brand name is covered in individual PHOTOFACTs because it is possible to do basic service work on similar chassis with just one PHOTOFACT, but all the necessary information may not be available. Important data pertaining to specific models and brand names may be missing:
1. There will not be a proper list of cabinet parts.
2. There will not be information on differences in the tuner section, component values, or added features and safety items.
3. There will not be accurate parts lists because manufacturers do change part numbers.

Ordering a PHOTOFACT set by chassis number alone can end up costing technicians time, money and perhaps some irritation. However, it is not a good idea to order only by model number, either. Some manufacturers keep the same model number for a set but use several different chassis in it over time. This can happen when a manufacturer contracts with an outside supplier for certain chassis needs. If this supplier changes, the chassis can also change. To insure that a technician gets the right information, a PHOTOFACT needs to be ordered using *both* the model and chassis numbers.

COILS & TRANSFORMERS

Item No.	Function/Rating	Mfr. Part No.
DY1	Yoke 90° Horiz 1.22mH Vert 15.4mH	4835 150 17102
FB401, 02	Ferrite Bead	4835 526 17002
FB431, 32	Ferrite Bead	4835 526 17002
L1	47µH	4835 157 57763
L200	VCO 45.75MHz	4835 157 57485
L204	AFT	4835 157 57594
L214	Sound Discriminator	4835 157 57113
L230	27µH	4835 157 57119
L301	3.3µH	4835 157 57154
L311	1.2µH	4835 157 67003
L312	12µH	4835 157 57048
L313, 14, 15	4.7µH	4835 157 67011
L357, 58, 59	27µH	4835 157 67019
L365	12µH	4835 157 57048
L386, 87	3.3µH	4835 157 57266
# L400	Line Filter	4835 152 17001
L405	1.8µH	4835 152 27029
L409	100µH	4835 157 57047
L422	2.2µH	4835 157 57752
L423	.68µH	4835 157 57751
L430	.7µH	4835 152 27036
L441	Ferrite Bead	4835 526 17009
L448	10µH	4835 152 27002
L452	2.2µH	4835 157 57752
L459	10µH	4835 157 57093
L460	42µH	4835 157 57063
# L499	Degaussing	4835 157 97072
L502	42µH	4835 157 57673
L503	Horizontal Linearity	4835 150 17101
L602	10µH	4835 150 57039
L605	15µH	4835 157 57756
L606	10µH	4835 150 57006
L610	10µH	4835 150 57039
L620	2.7µH	4835 157 57098
L625	10µH	4835 150 57004
L635	18µH	4835 157 67031
L636, 37, 38	1.2µH	4835 157 67003
L704	160mH	4835 157 57657
L900	15µF	-
# T401	Power	4835 148 87286
# T450	Standby	4835 148 87251
T501	Horizontal Drive	4835 142 47018
# T502 (1)	Horizontal Output	4835 140 67088
V421 (L421)	.7µH	4835 152 27036
V546	5.6µH	4835 152 27038

For SAFETY use only equivalent replacement part.
(1) Focus and screen controls are part of T502.

COILS & TRANSFORMERS continued

Item No.	Function/Rating	Mfr. Part No.
PIP BOARD		
L101	1.2µH	4835 157 67028
L104	6.8µH	4835 157 57061
L105	.47µH	4835 157 57014
L107	6.8µH	4835 157 57061
L201	100µH	4835 157 57818
L202	6.8µH	4835 157 57061
L301	.47µH	4835 150 57045
L302	39µH	4835 157 67021
L303	3.3µH	4835 157 57832
L304	6.8µH	4835 157 57061
L305	10µH	4835 150 57008
L306	12µH	4835 157 57831
L307	5.6µH	4835 157 57833

CABINET PARTS

Item	Mfr. Part No.
Button Assembly	4835 219 47231
Cabinet Back Cover	4835 432 97396
Cabinet Front	0014 688 90014
Crystal Bezel	4835 450 67136
Grille	4835 459 47048
Lens, IR	4835 381 17006
Nameplate	4835 459 17388

Figure B-1. Examples of PHOTOFACT parts lists.

When ordering a PHOTOFACT, it is also important to be sure to use the complete numbers. Some manufacturers will add a suffix code to the model number. This suffix code is usually stamped below the model number and will read SUFFIX, followed by a letter of the alphabet. These suffix codes

APPENDIX B: PHOTOFACT

are very important because they signify differences between models. These differences could be minor or major. A technician with a TV with a suffix not listed in a PHOTOFACT should call Howard W. Sams' customer service at 1-800-428-7267 to see if there is any additional information on the set. Customer service may be able to determine if the differences between the TV and an existing PHOTOFACT is major or minor. Also, some manufacturers include a production run number along with the chassis number. Like suffix numbers, production run numbers indicate differences between chassis. Howard W. Sams lists production run numbers that have been covered by indicating RUN, followed by the appropriate number sequence.

In every PHOTOFACT is a list of the parts and components for the model covered to help the technician select the proper replacement and repair the unit quickly and efficiently. (*Figure B-1.*) Surveyed technicians say that the parts list is one of the main reasons they buy PHOTOFACT.

Every single part and component found in the set is not necessarily listed. Instead, the focus is placed on the parts that meet general servicing needs so that the technician doesn't have to hunt through volumes of information that aren't likely to be used. The only items that are listed include:
1. Special or unusual: parts that technicians don't normally keep on hand, such as cabinet parts or very low wattage resistors.
2. Safety items: parts that legally require exact replacements. All safety items are shaded on the schematic and marked with a # on the parts list as a reminder.
3. Substitutions: items that can be replaced with equivalent parts from participating electronics vendors.

For parts that are not listed, values are shown on the schematic so that technicians are never without the service information they need. There also could be alternate values listed on the schematic. It is always a good idea to look at the component to determine its value before replacing it. Parts not listed are commonly available from local electronics distributors or from the technician's own parts bin. For example, if a set needs a 1000 ohm, 1/2W resistor, it is a lot faster and cheaper to get it from a local distributor than to order it from the manufacturer.

All semiconductors in a set are covered in PHOTOFACT, including those associated with the tuner when that information is available from the manufacturer. As many different alternatives as possible is given for each semiconductor. A recommended replacement is the best one available at the time the PHOTOFACT was produced.

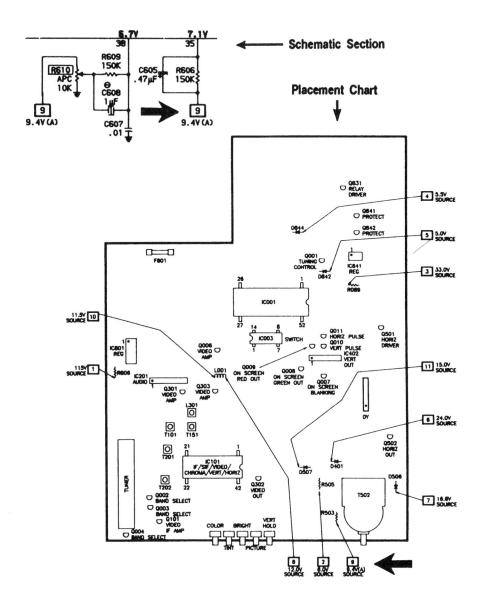

Figure B-2. A PHOTOFACT placement chart.

The replacement manufacturers directly supply all recommended replacement parts listed in PHOTOFACT. When ordering parts, always state the model number, part number, and part description. Howard W. Sams' technicians do their best to include everything a technician needs to order the correct parts, but please keep in mind that the manufacturer does not always supply a number, so it may be necessary on occasion to use some guesswork.

PHOTOFACT also includes a placement chart to simplify pinpointing and checking problem locations. The placement chart does not show every component on the board—only the physical positions of semiconductor devices, sources, and test points with safety precautions. (*Figure B-2.*) In a

APPENDIX B: PHOTOFACT

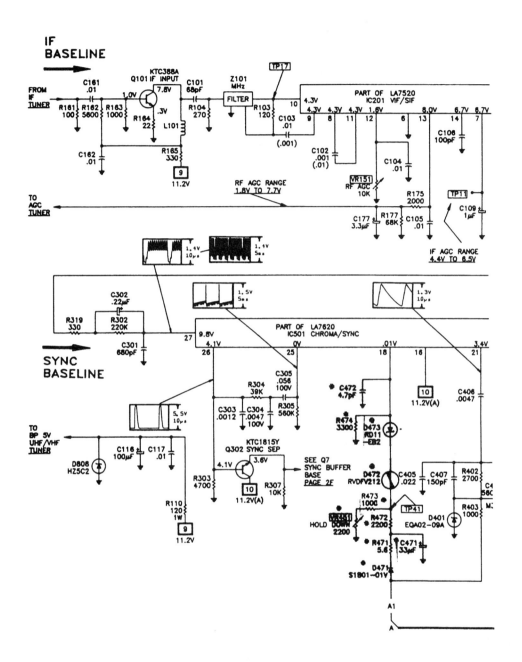

Figure B-3. A PHOTOFACT schematic showing the IF and sync baselines.

problem area, the source can easily be identified, and the voltage can be checked without tracing it all the way through the circuit or returning to the schematic. All PHOTOFACT placement charts are drawn on CAD.

All PHOTOFACT schematics are laid out with a major signal progression running from left to right. If a straight edge is laid on the schematics, three

major baselines and one or more minor baselines can be seen. The three major baselines represent:
1. The signal from the tuner to the CRT's picture and chroma sections.
2. The horizontal and vertical syncs to the yoke.
3. The horizontal line to the yoke and the flyback.

A minor baseline can start out as audio and then become a major baseline if SAP or stereo functions are included in the schematic. (*Figure B-3*.)

After making a visual inspection of the unit and not being able to localize the problem, the technician should make certain that the power supply is working. An inoperative or defective power supply can make further troubleshooting very difficult. The AC-DC source points and the flyback-generated source need to be confirmed and measured properly. (The 120V AC line-in is especially important.)

PHOTOFACT schematics show the AC-DC source points and the flyback-generated source in two main areas. One area is near the AC input. This

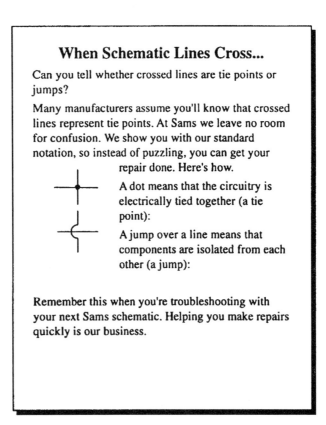

Figure B-4. *A PHOTOFACT tip describing crossed schematic lines.*

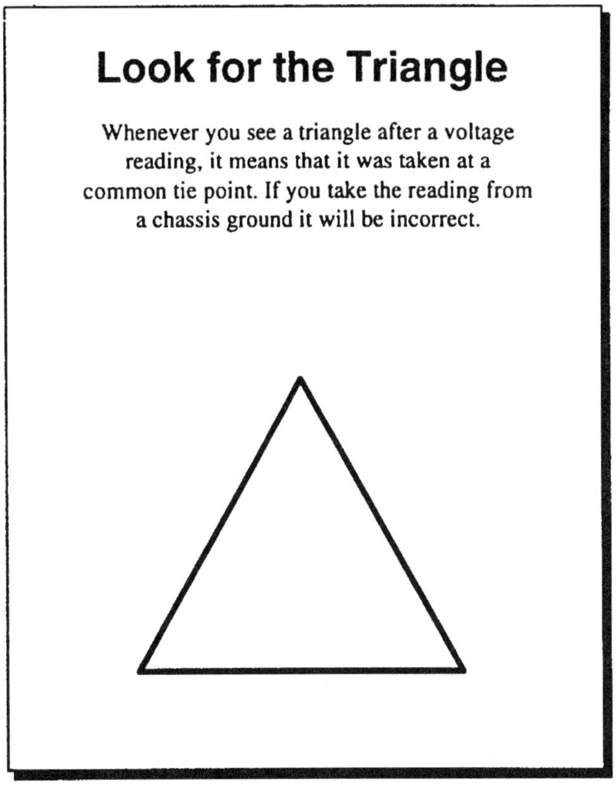

APPENDIX B: PHOTOFACT

Look for the Triangle

Whenever you see a triangle after a voltage reading, it means that it was taken at a common tie point. If you take the reading from a chassis ground it will be incorrect.

Figure B-5. A PHOTOFACT tip on voltage readings.

reflects the voltages developed from the line voltage. The other source comes from the flyback collapsing field and is located either near the flyback or below the AC power supply.

After the major power source has been affirmed, a good signal needs to be injected at the left side of the schematic tuner or tuner output, and the schematic baseline should be divided at midpoint. An oscilloscope can be used to confirm a good signal. If there is a good signal through the midpoint, divide the remaining right section in half. Continue to run the signal through this portion of the schematic circuitry using an oscilloscope.

If a bad signal is received, keep dividing the schematic down, running the signal from left to right until the problem has been pinpointed. Remember that start-up problems with a variable AC power supply allow the technician to test suspect areas by starting at a low voltage (60V AC, for example), and slowly increasing the voltage.

Every PHOTOFACT schematic follows an easy-to-use service format. *Figures B-4* and *B-5* outline a couple of things service technicians should watch for when reading a PHOTOFACT schematic.

APPENDIX B: PHOTOFACT

Quiz

1. *True or False*—All Sams schematics are laid out with a major signal progression running from left to right across the page.
2. *True or False*—There are three major baselines on PHOTOFACT schematics: the signal through the picture and chroma sections to the CRT; the horizontal and vertical sync through the sync section and vertical output to the yoke; the signal through the horizontal output to the yoke and the flyback.
3. *True or False*—Always test suspect areas with the set connected directly to the AC power supply.
4. *True or False*—To reduce the schematic size, cut it into handy smaller sections.
5. *True or False*—The PHOTOFACT parts list does not list every single part and component in the set.
6. *True or False*—All safety items are marked with a 0 on the schematic.
7. *True or False*—A value shown in parenthesis () on the schematic indicates a missing component.
8. *True or False*—A triangle on the schematic indicates that the voltage reading was taken from a chassis ground.
9. *True or False*—PHOTOFACT takes voltages with no signal because they're the easiest to reproduce.
10. *True or False*—All you need to look at in a PHOTOFACT is the schematic.
11. *True or False*—The warning DO NOT MEASURE means that you don't have to test a particular component.
12. *True or False*—A missing arrowhead on a photo identifier indicates a hidden component.
13. *True or False*—You can substitute replacements for safety items.

Key

1. True. This makes it as easy as possible for you to trace signals and troubleshoot the set.
2. True.
3. False. You should use a variable AC power supply transformer to gradually increase the power.
4. False. Simply fold the schematic along its long dimension, or vertically through the center to reduce the schematic to either four or two pages.
5. True. Parts that are commonly available from local electronics distributors are not listed. However, the part value is provided on the schematic.
6. False. Safety items are marked with a # and are shaded on the schematic.
7. False. Parenthesis () indicate an alternate value for a component.
8. False. A triangle on the schematic indicates a common tie point.
9. True. There are so many different signal generators available that Sams could not reproduce every signal.
10. False. Always read all notes, charts and symbols carefully. They are full of important information.
11. False. It is a warning that a particular component may fail when you are taking voltages, or that the voltage present at that test point could damage your test equipment.
12. True. The component could be underneath a shield or on the bottom of the board.
13. False. A safety component must be replaced with an equivalent part rated for safety and performance.

Bibliography

Davidson, Homer L. <u>Troubleshooting and Repairing Solid-State TVs</u>, Second Edition. New York: TAB Books, McGraw-Hill, Inc., 1992.

Deane, Leslie D. & Young, Calvin C., Jr. <u>TV Servicing Guide</u>. Indianapolis, IN: Howard W. Sams & Co., 1981.

Gottlieb, Irving M. <u>Test Procedures for Basic Electronics</u>. Indianapolis, IN: PROMPT Publications, Howard W. Sams & Co., 1994.

Hollomon, James K., Jr. <u>Surface-Mount Technology for PC Boards</u>. Indianapolis, IN: PROMPT Publications, Howard W. Sams & Co., 1995.

Howard W. Sams Editorial Staff. <u>PHOTOFACT Television Course</u>, Fifth Edition. Indianapolis, IN: Howard W. Sams & Co., 1987.

Johnson, J. Richard. <u>Schematic Diagrams</u>. Indianapolis, IN: PROMPT Publications, Howard W. Sams & Co., 1994.

Lenk, John D. <u>Lenk's Digital Handbook</u>. New York: McGraw-Hill, Inc., 1993.

Lenk, John D. <u>Lenk's Television Handbook</u>. New York: McGraw-Hill, Inc., 1995.

Lines, David. <u>Power Supplies</u>. Indianapolis, IN: PROMPT Publications, Howard W. Sams & Co., 1992.

Prentiss, Stan. <u>Modern Television Service and Repair</u>. Englewood Cliffs, NJ: Prentice Hall, Inc., 1989.

Sloop, Joseph G. <u>Advanced Color Television Servicing</u>. Indianapolis, IN: Howard W. Sams & Co., 1985.

Sloop, Joseph G. <u>Television Servicing With Basic Electronics</u>. Indianapolis, IN: Howard W. Sams & Co., 1981.

Tinnell, Richard W. <u>Television Symptom Diagnosis</u>, Second Edition. Indianapolis, IN: Howard W. Sams & Co., 1985.

Van Valkenburgh, Nooger & Neville, Inc. <u>Basic Electricity</u>. Indianapolis, IN: PROMPT Publications, Howard W. Sams & Co., 1992.

Van Valkenburgh, Nooger & Neville, Inc. Basic Solid-State Electronics. Indianapolis, IN: PROMPT Publications, Howard W. Sams & Co., 1992.

Zwick, George, & Davidson, Homer L. TV Repair for Beginners, Fourth Edition. New York: TAB Books, McGraw-Hill, Inc., 1991.

Index

A

A/V inputs 92
A/V switching circuits 92, 101, 110, 114
AC 59, 64, 76
AC cord 27
AC current 173, 174, 182
AC current draw 77
AC input 82, 152, 155
AC input voltage 57, 77
AC line input 66
AC outlet 27
AC output voltage 57
AC output voltage value 59
AC plug 21, 22, 27
AC power 182
AC power cord 21
AC signal 95
AC source 73, 75, 168
AC voltage 22, 57, 59, 60
AC voltmeter 21, 27
AC waveforms 33
Active components 39
Adjustable coil 92
Adjustments 36
AFT 166, 167
AFT (automatic fine tuning) circuit 164, 166, 167
AFT correction voltage 165
AGC (automatic gain control) 12, 85, 164
AGC circuits 116
AKB IC 98
Alarm clock 177
Alignment 94
Alternate fields 139, 140
Alternation 62, 63, 64
Alternation B 62
AM frequencies 113, 114
AM frequency carrier 114
AM signals 9, 13
AM video carrier 162
Ammeter current 77
Amplification 64, 97
Amplified signal 13, 87

Amplifier 10, 39, 95, 114, 127, 131, 139, 140, 141
Amplifier stage 127, 128
Amplitude 39, 140
Amplitude variations 57
Amps 61
Analog 33, 93
Analog control 92, 93
Anode 61, 62, 70, 105, 106
Anode connection 25, 151
Anode lead 22
Anode terminals 104, 153
Anode voltage 60, 105, 155
Anode-cathode junction 61
Antenna 8, 10, 17, 45, 90, 162, 163, 164
Antenna connection 167
Antenna lead 22, 168
Antenna screw heads 27
Antenna signals 11
Antenna terminals 21
APC 15, 127, 128, 129, 130, 131
APC (automatic phase control) circuit 127, 129
Aperture 102
Arc 152
Arcing 36, 47, 82, 102, 145, 152, 188
Aspect ratio 102
Assigned frequency 11
Audio 6, 9, 10, 12, 13, 14, 85, 86, 89, 90, 91, 92, 101, 112, 175
Audio amplifier 37, 115, 116, 117, 118
Audio amplifier stage 14, 117
Audio amplifier/control circuits 114
Audio carrier 165
Audio circuits 112, 116, 172
Audio controls 175
Audio detector 114
Audio distortion 37
Audio frequency 120
Audio IF signals 87, 113
Audio IF stage 113
Audio level 14
Audio matrix 114, 115
Audio mode circuits 175
Audio power amplifier 118
Audio processing circuits 112, 114, 116
Audio processing stages 112, 113

Audio processing system 14, 112
Audio processor 116
Audio product 36
Audio signal 10, 14, 91, 92, 112, 114, 115, 117, 120, 179, 203
Audio signal generator 118
Audio stages 119
Audio symptoms 202
Audio system 116
Audio trap 118
Audio trap circuit 91
Audio/video (A/V) switching circuit 92
Auto channel programming 175, 177
Auto programming sequence 176
Automatic brightness control 96
Automatic brightness limiter (ABL) circuit 157
Automatic chroma control (ACC) 99
Automatic gain control (AGC) 12, 17
Automatic kine bias (AKB) 98
Automatic phase control (APC) 15, 129
Automatic picture-level control 173
Average current 43

B

Background noise 203
Balance 175
Band selection 173
Bandpass filter 92, 114
Bandwidths 13
Bare wires 48
Base 38, 39, 40, 77, 132, 158, 175
Base bias voltage 91
Base circuit 77
Base current 72
Base resistors 120
Base signal 39
Base terminal 95, 119
Base voltage 40, 77
Base-collector junction 40
Basic electronics 48
Basic equipment 32
Basing 38
Bass 175
Batteries 182
Beam current 98

Beams 106
Bi-metal holders 102
Bias 40, 91, 120, 129
Bias current 98
Black bars 66, 194
Black box diagram 10
Black raster 179
Black-and-white information (luminance) 97
Blank 140
Blanking bar 128, 194, 201
Blanking circuit 8, 179, 182
Blanking signal 139, 143
Blatting 119
Bleeder resistor 155
Block diagram 9
Block diagram 9, 10, 56, 73, 75, 84, 101
Blooming 104, 137, 157, 194
Blooming—white raster 104
Blotch 105
Blue 8, 97, 105
Blue beams 107
Blue guns 106
Blue signals 179
Blue video amplifiers 97
Board 44, 153, 174
Board wiring 27
Boost 75, 81, 151, 154, 157, 165, 168
Boost fuse 137
Boost fuse blows 188
Boost voltage 154
Bottom-to-top instructions 7
Brand 21, 25, 36, 46, 101, 125, 173, 175, 178, 182
Breathing picture 66, 194
Bridge 41, 47
Bridge network 63
Bridge rectifier 63, 72
Brightness 8, 105, 137, 144, 152, 157, 180, 197
Brightness control 96
Brightness levels 22
Brightness limiter 96
Brightness problems 195
Buck 75
Buck regulator 75
Buffer 92
Burn 24
Burn marks 81
Burned areas 82

Burned components 76, 157
Burned connections 151
Buzz 114, 116, 203
Bypass capacitor 40

C

Cabinet 20
Cabinet parts 21
Cable 10
Cable access signals 11
Cable stations 90
Calibration 22
Camera 7, 8, 97
Capacitance 45, 165
Capacitance arrays 45
Capacitance tester 66
Capacitive reactance 65
Capacitor 21, 27, 40, 41, 65, 66, 68, 74, 91, 118, 119, 132, 133, 134, 136, 140, 143, 145, 154, 165, 166, 174
Capacitor lead 118
Capacitors 22, 33, 39, 40, 64, 65, 66, 67, 68, 118, 134, 145
Carriers 162
Catch diode 75
Cathode 43, 62, 70, 103, 104
Cathode band 71
Cathode drivers 98
Cathode ray tube 7
Cathode terminals 104
Cathode voltage 60
Center channel 115
Center speaker 115
Center tap 63
Ceramic filter 177
Channel 10, 11, 86, 87, 115, 162, 163, 165, 166, 167, 173, 175, 176, 177, 193
Channel frequency 164, 166
Channel information 176
Channel memory 172, 176, 190
Channel memory (RAM) 175
Channel numbers 176
Channel tuning 177
Character generator 178, 179
Characters 178
Charge storage 65
Chart format 9

Chassis 22, 36, 39, 44, 56, 67, 78, 81, 82, 93, 98, 102, 104, 105, 112, 119, 120, 125, 135, 151, 153, 154, 157, 162, 163, 165, 168, 173, 179
Chassis ground 44
Chassis shutdown 188
Chip passives 45
Choke 66, 68
Chopper 72
Chroma 93, 100
Chroma processing 14, 97
Chroma signal information 92
Circuit 10, 20, 26, 36, 37, 39, 41, 43, 48, 56, 59, 61, 65, 67, 70, 71, 72, 75, 76, 77, 78, 82, 85, 87, 91, 97, 98, 102, 104, 116, 117, 118, 119, 120, 125, 129, 130, 131, 134, 137, 143, 145, 150, 151, 157, 158, 168, 173, 174, 175, 178, 179, 182
Circuit board 20, 27, 49
Circuit breaker 24
Circuit diagram 163
Circuit ground 59
Circuit performance 45
Circuit resistance 41
Circuit voltage 37, 40
Circuitry 67, 89, 98
Class A amplifiers 39
Class B amplifiers 39
Class C amplifiers 39
Click 117
Clip wire 118
Clock 172, 176, 177
Clock circuit 177, 178
Clock functions 177
Clock line 177, 178
Closed caption 179
Closed caption circuit 179
Closed caption decoder circuit 179
Closed caption information 178, 179, 180
Closed loop 72, 130
Coil 39, 40, 58, 117, 120, 132, 166
Cold current leakage 21, 22, 27
Cold ground 56
Cold leakage checks 27

INDEX

Cold solder joint 42, 44
Cold solders 120
Cold spray 44, 119, 120
Collector
   38, 39, 40, 77, 78, 118, 134
Collector current 72
Collector lead 39
Collector terminals 56
Collector voltage 36, 91
Collector-emitter voltage 72
Color
   8, 16, 89, 90, 95, 97, 99, 168, 180
Color amplifier 97, 99
Color bar 34
Color bar generator (NTSC) 32, 34
Color bleeding 195
Color burst 97, 100
Color carrier 12, 165
Color circuits 98
Color controls 99
Color CRTs 35
Color demodulator circuits 97, 99
Color fringes 106
Color guns 99, 105
Color IF amplifier 97, 98, 99
Color intensity controls 98
Color killer 97, 98, 99
Color lead 153
Color loss 99
Color modulated signal 97
Color oscillator 92, 100
Color oscillator frequency 99
Color oscillator output signal 99
Color picture 97
Color picture tube jig 104
Color processing 14
Color processing circuits 97, 100
Color setup procedures 34
Color shifts 100
Color signal 92, 97, 99, 125
Color signal demodulator 99
Color signals
   14, 98, 100, 102, 112
Color sync 99
Color sync amplifier 98, 100
Color sync signal 97
Color synchronization 97
Color televisions 14, 26
Color video amplifier 99
Comb filter 92, 109
Commands 179
Common collector 39

Common collector amplifiers 39
Component callout 49
Component failures 39
Components 22, 23, 27, 36, 37,
   38, 40, 41, 42, 44, 45, 46,
   47, 48, 50, 73, 77, 81, 86,
   95, 100, 105, 116, 117, 118,
   119, 120, 125, 129, 130,
   131, 132, 133, 134, 140,
   145, 150, 151, 153, 154,
   155, 157, 163, 167, 173, 174
Composite audio signal 114, 116
Composite video 13
Composite video signal
   13, 89, 109, 125
Concert hall 115
Conducting material 24
Conducting paths 64
Conduction 40, 63, 75, 140
Conductive coating 102
Conductive plates 65
Cone 117, 118
Configuration 125, 173
Connections 20, 36, 42, 46, 47,
   56, 81, 82, 105, 116, 117,
   129, 130, 140, 145, 151, 152,
   153, 154, 155, 157, 168
Consumers 20, 154
Contact points 46
Contacts 179
Contrast
   8, 95, 105, 167, 180, 197
Contrast control 13, 95
Control 39
Control element 69, 72, 73, 74
Control element switch 75
Control knobs 20, 27
Control shafts 21, 27
Control voltage 12
Controls 8, 36, 92, 172
Conventional current flow 43
Conventional power supplies
   56, 66
Convergence 106, 107
Convergence adjustments 106
Convergence magnet nut lock 107
Convergence procedures 107
Convergence yoke 106
Conversion 75
Converting signals 9, 10
Core 76
Core material 76

Core saturation 151
Corona 23
Correction voltage 129, 131
Countdown circuit 131, 140
Counter voltage 58
Covers 20, 27
Cracked traces 174
Crackling 119, 202
Cracks 25
Cross-hatched patterns 34
Crosshatch pattern 107
CRT 7, 13, 15, 16, 22,
   23, 24, 25, 26, 33, 35, 37,
   78, 84, 94, 96, 99, 102, 103,
   104, 105, 106, 107, 125, 133,
   134, 139, 143, 144, 151, 153,
   155, 157, 178, 179, 180
CRT anode lead 26
CRT connections 153
CRT screen 125
CRT test jig 32, 33
CRT tester 32, 33, 104
Current 33, 40, 42, 43, 44,
   58, 59, 60, 61, 63,
   64, 70, 73, 76, 91,
   96, 135, 136
Current check 27
Current demand 68
Current overload 68
Current resistors 39
Current-limiting resistor 71
Customer 20, 36
Customer preferences 173, 174
Customized settings 172, 173
Cutoff 40

D

Damper diode
   133, 134, 136, 137, 138
Data 37
Data line 177
Data signals 178, 180
Data transfer 93
DC 59, 64, 93
DC bias 96
DC input voltage 73
DC line input 66
DC output voltage 59
DC power supplies 65
DC source 73
DC value 64

DC voltage 33, 56, 57, 60, 62, 63, 65, 68, 77, 78, 82, 93
Decode 114
Defective diodes 43
Defective ICs 44
Defective SCRs 43
Defective stage 39
Defective transistors 42
Deflection circuit diagram 125
Deflection circuits 124, 125, 132
Deflection coil 137
Deflection signal 131
Deflection yoke 15, 133, 135, 143, 145
Deflection yoke coil 134, 137
Deflective 153
Degaussing coil 26, 32, 105, 106
Delay 173
Delay line 92, 109
Delay time 173
Demagnetize 106
Demodulate 87, 114
Demodulator 115, 116
Demodulator circuit 97, 116
Depletion 70
Desoldering iron 46
Detector 114
Detector coil 116, 118, 120
Diagnostic tool 35
Dielectric 65
Digital 33
Digital circuits 48
Digital control 92, 93
Digital multimeter (DMM) 32, 33
Digital sync 166
Digital tuning 166
Dim color 196
Dim picture 202
Diode 33, 41, 42, 43, 56, 60, 61, 62, 63, 66, 67, 68, 70, 71, 74, 75, 78, 88, 130, 145, 154, 165, 173
Diode function 67
Diode test function 40
Dips 68
Discharge 22, 26, 140
Discrete tuner 163
Dissipation 68
Distorted sound 116, 119, 202, 205
Distortion 37, 136, 144, 145, 175
DMM 40, 43, 44, 66, 67, 118, 132, 150, 154, 158
Documentation 35, 101, 173
Dolby ProLogic™ 115
Dolby™ processing 115
Dolby™ Surround sound 114, 115
Dot 34
Double-sided boards 44
Downward tearing 131
Dressed leads 20
Drift 166, 167, 177, 194
Drifting frequencies 167
Driver amplifier 118
Dynamic convergence 105, 106
Dynamic misconvergence 106

E

Earth ground 21, 22, 27
Electrical audio signal 14
Electrical charges 65
Electrical circuits 22
Electrical components 23, 27
Electrical connection 59
Electrical current 41
Electrical shocks 22
Electrical signals 6
Electrical system 56
Electricity 60
Electrode 103
Electrolytic capacitor 39, 118
Electrolytics 145
Electromagnetic interference (EMI) 70
Electron beam deflection 132
Electron beams 15, 16, 102, 103, 105, 124, 125, 135, 139
Electron emitter 103
Electron guns 7, 15, 16, 96, 97, 98, 102, 104, 157
Electronic circuits 64
Electronic components 32
Electronic equipment 32, 57
Electronic valve 61
Electronics books 48
Electrons 60, 65, 70, 103
Electrostatically sensitive parts 20, 26
Emitter 38, 39, 40, 56, 119, 134
Emitter circuit 39, 95
Emitter resistors 40, 118
Emitter terminal 36
Emitter-collector junction 42
Emitter-follower 39
Emitter-to-collector short 91
Energy 56, 59, 74, 78
Energy storage 73
Energy transfer 59
Equipment 35
Erratic responses 132
Erratic sound 120
Error amplifier 69, 72, 74
Exact replacement 46, 78, 88, 138, 153, 168
Excessive heat 39, 46
Excessive high voltage 27
Excessive voltage 70, 105, 153, 154
External bias 145
External load 67

F

Factory 20
Factory replacement 45
Factory settings 173, 174, 175, 177, 179
Factory specifications 21
Farad 66
Faulty filter capacitor 66
Faulty IF amplifier 167
Faulty resistor 132
Faulty solder connections 174
Faulty tuners 166
Features 173
Federal regulations 154
Feedback 131, 140, 141
Feedback circuit 77
Feedback loop 131
Feedback regulator 72
Ferrite 76
FET (field effect transistor) devices 47, 75, 164, 166
Field 7, 59
Filter capacitors 22, 47, 64, 66
Filter circuit 64, 73
Filter regulator 119
Filtering 56, 57, 64, 68, 70, 75, 76, 114, 115, 117
Final video amplifier circuit 105
Fine tuning 164, 166, 172
Fine-tune 165
Fire hazards 24
Fires 23
First anode 103
First stage 90
First video amplifier 110

INDEX

Fixed delay 177
Fixed frequency reference oscillator 177
Fixed radio frequency amplifier 113
Fixed repetition pattern 177
Fixed voltage 70
Flammable materials 24
Flaws 25
Flickering 7, 16
Flip-flop circuits 140
Floor 22
Fluorescent coated screen 103
Flyback 82, 133, 142, 151, 152, 153, 155, 157
Flyback circuit 105, 154, 155
Flyback regulator 75
Flyback transformer 22, 36, 151, 152, 153, 156
Flyback transformer stage 16
Flyback winding 154
FM audio signal 14, 114
FM radio signals 113
FM signals 9, 13, 14, 113
FM sound carrier 162
Focus 105, 137, 153, 157, 197
Focus circuits 155
Focus control 105, 152, 153
Focus problems 153, 200
Focus voltage 105, 155
Folding picture 199
Foreign particles 20
Forward bias 40, 60
Forward current 43
Forward diode drops 64
Forward-Biased Diode (VF) 61
Frame 7
Frayed wires 23
Freeze 101
Frequency 11, 64, 78, 113, 117, 129, 131, 132, 139, 162, 164, 165, 177, 178
Frequency (channel) drift 188
Frequency band switch decoder 166
Frequency comparison circuit 177
Frequency converters 165
Frequency counter 178
Frequency lock 173
Frequency output 132
Frequency selectivity 163

Frequency synthesis (FS) tuner 166
Frequency synthesizer 173
Frequency translators 165
Frequency variations 118
Front panel controls 172, 188
Front-end control 173
FS circuit 166
Full load 67
Full-wave rectifier 63, 64
Function (command) 39, 182
Fuses 24, 36, 66, 67, 81, 137, 158

G

Gain 89, 98
General operating guidelines 20
General techniques 35
General television problems 188
Generic tuner circuit 163
Germanium 37, 39
Germanium diode 61
Glass 24
Gloves 25
Glue 46
Goggles 26
Green 8, 97, 105
Green color amplifier 99
Green electron gun 107
Green screen 107
Green signals 179
Green vertical band 107
Grid circuits 104
Grid map 49
Ground 20, 22, 25, 27, 36, 39, 63, 94, 118, 175
Ground lead 151
Grounding wrist strap 23, 166
Guns 106

H

Hair dryer 46
Half-cycle alternations 62
Half-wave rectifier 61, 67
Handle brackets 21, 27
Hardware 21
Heat 40, 46
Heat sink 47, 143
Heated oxide surfaces 103
Heater 46
Heater element 105

Henries 59
Herringbone pattern 91
Heterodyne detectors 165
Heterodyning 162, 163
High frequency 115
High frequency signals 94
High voltage 22, 23, 26, 35, 39, 104, 105, 125, 143, 151, 154, 155
High voltage arcing 47
High voltage circuits 26
High voltage condition 138
High voltage level 153
High voltage meter 26
High voltage power 133
High voltage probe 32, 35, 85
High voltage socket 102
High-energy magnetic fields 46
High-frequency circuits 37
High-pitched squeal 193
High-voltage arcing 152
High-voltage bleeder resistors 155
High-voltage buildup 151
High-voltage cable 105
High-voltage circuit 22, 150, 151, 153, 154, 157
High-voltage connections 20
High-voltage diodes 151
High-voltage limits 23
High-voltage meter 22, 23
High-voltage power supplies 16, 150
High-voltage probe 22, 105, 144, 150, 151, 153, 158
High-voltage problems 157
High-voltage protection circuit 154, 155
High-voltage rectifier 23
High-voltage regulation 154, 157
High-voltage shutdown 153
High-voltage stages 145
High-voltage values 23
High/low voltage 173
Highest resistance 75
Hissing 86, 87, 118, 167, 202
Hold-down circuit 154
Horizontal bar 71
Horizontal blanking 96
Horizontal circuits 82, 119, 158, 178, 182
Horizontal control 131

Horizontal deflection (drive) circuits
102, 124, 132
Horizontal deflection circuit IC 131
Horizontal deflection circuits
124, 126, 132, 145
Horizontal deflection IC
129, 130, 132
Horizontal differentiator 126, 127
Horizontal hold 127
Horizontal hold control
15, 127, 129, 131
Horizontal lines 7, 107, 133, 139
Horizontal oscillator
15, 100, 129, 130, 131, 132, 140
Horizontal oscillator circuit 131
Horizontal oscillator frequency 99
Horizontal output 15, 78
Horizontal output circuit
79, 125, 131, 133, 134, 137,
138, 151, 155, 158
Horizontal output collector 78
Horizontal output signal 81
Horizontal output stage 15
Horizontal output transformer 152
Horizontal output transistor
22, 76, 134, 136
Horizontal output voltage 78
Horizontal pin cushion effect 141
Horizontal pulse 140
Horizontal rate 140
Horizontal scan
15, 16, 124, 133, 140, 200
Horizontal scan lines 127
Horizontal scan signal 15
Horizontal scanning 124, 125,
134, 139, 140
Horizontal signals
124, 125, 127, 128, 145
Horizontal stability 131
Horizontal stages 36
Horizontal sync 198
Horizontal sync circuits 178, 180
Horizontal sync processing circuits
138
Horizontal sync pulse 129, 131
Horizontal sync signal 14,
15, 126, 132
Horizontal sync signal rate 15
Horizontal synchronizing pulses 96
Horizontal tearing 7,
15, 66, 127, 128
Horizontal waveform 135

Horizontal white line 144, 196
Horizontal-sync signals 7
Horizontally narrow picture 200
Horizontally unstable picture 66
Hot areas 36
Hot chassis 66
Hot connections 24
Hot current leakage 22
Hot ground 56
Hot leakage 27
Hot leakage current 21, 27
Hum 36, 64, 66, 87, 89, 116,
117, 118, 119, 203
Hybrid tuners 163

I

IC 36, 44, 46, 67, 68, 85, 86,
87, 88, 92, 93, 94, 100, 112,
117, 118, 119, 125, 129,
130, 131, 138, 139, 140,
145, 150, 166, 167, 172,
173, 175, 177, 178, 180,
182, 183
IC board 173
IC pins 118
IC processors 46
IC regulators 68, 70
IC terminal voltages 113
IF 10
IF (intermediate frequency) amplifier 11, 12, 13, 84, 85, 86, 87,
88, 89, 166, 167
IF amplifier circuit 87, 109
IF circuit 37, 177
IF circuit lead 168
IF frequency 12
IF output 166
IF output signal 89
IF processing circuit 116
IF processing stages 112
IF signal 12, 162, 165, 166, 168
IF signal waveform 12
IMC (insertion-mount components)
44
Impedance probe 178
Implosion 22, 24, 26
Incorrect color 99, 199
Incorrect colors 99, 194
Independent stage 10
Inductance 45, 58, 76
Inductive reactance 59, 64
Inductor 64, 67, 73, 74, 75, 76

Infinity 21, 27
Information 173
Information displays 173
Infrared card 182
Infrared circuits 179
Infrared receiver 180
Infrared receiver circuits 180
Infrared signals 180
Injected signals 37
Inner board wiring 20
Input 10, 36, 37, 39, 67,
76, 77, 89, 93,
96, 98, 116, 117, 129, 130,
131, 132, 135,
153, 155, 158, 162, 179
Input connections 119
Input filter 73, 76
Input primary voltage 62
Input signal 88, 99, 138, 163
Input stage 102
Input values 140
Input voltage
57, 58, 68, 69, 72, 73,
75, 91, 130, 132, 144
Instructions 22, 26, 168
Insulation 20, 40, 152
Insulators 20, 22, 27, 65
Integrated circuit (IC) 13, 20, 40
(See IC)
Integrated transformer 151
Intense color 98, 195
Interference 95, 116
Interlaced scanning 7, 139, 140
Intermediate frequency (IF) 10,
163 (See IF)
Intermediate-frequency (IF) amplifiers 10
Intermittent color 98
Intermittent focus 105
Intermittent scanning raster 104
Intermittent sound 116, 119,
202, 205
Intermittent sound problems 120
Internal series-pass transistors 68
Inverter regulator 75
Ions 65
Iron core 58, 59
Iron laminated cores 76
Isolated ground 21, 27
Isolation 59
Isolation material 20

INDEX

Isolation transformer
21, 22, 26, 27, 56, 66,
73, 76, 150

J

Jittery picture 129, 130
Joints 46, 48
Jump 24
Jumper 21, 27, 47
Jumper wires 47
Junction 41

K

Keyboard (front panel control)
172, 173, 175, 179, 182, 188
Keyboard buttons 179
Keyboard circuit 182
Keystoning 137, 145, 196
Kine board 96, 99, 105

L

L-R demodulator circuit 117
L-R signal 115
L-R stereo signal 115
Last video amplifier (LVA) 96, 116
(See LVA)
Lead dress 27
Leads 20, 22, 39, 42, 45,
46, 47, 48, 154
Leaky capacitor 41
Leaky component 56, 118, 157,
173
Leaky diodes 82
Leaky spark gap assembly 105
Leaky transistor 42, 117, 118
Left and right signals combined
(L+R) 114
Left channel 115
Left signal 114
Letters 178
Light raster 179
Line cord 22, 26
Line pairing 144, 145, 197
Line splitting 144, 145, 197
Line voltage 57
Linear 141
Linear conduction 141
Linear operating state 74
Linear regulated power supply 73
Linear signal 140
Linear supply 74

Linear system 73, 74
Linearity controls 16
Linearity correction 141
Linearity problem 145
Lines 152
Lithium batteries 176
Load 57, 63, 64, 69, 76, 79
Load circuit 70
Load current 63, 69, 72
Location information 16
Loop 130
Loose wiring 120
Lost color 199
Low bandpass filters 117
Low current applications 43, 61
Low power supplies 66
Low resistance 42
Low voltage capacitors 40
Low voltage measurement 82
Low voltage values 82, 157
Low wattage resistors 40
Low-pass filters 125
Low-power ohms range 41
Low-voltage windings 151
Low-wattage soldering iron 154
Lowest resistance 75
Luminance 14, 16,
92, 99, 102, 110
Luminance circuit 105
Luminance information 92
Luminance waveform 92, 93
LVA (low voltage amplifier) 96

M

Magnetic coils 103
Magnetic coupling 59
Magnetic energy 135
Magnetic field 46,
47, 58, 59, 75, 135, 139
Magnets 106
Main screen 101
Main stages 139
Manual degaussing coil 35
Manuals 9
Manufacturer
20, 25, 26, 36, 45, 46, 93,
125, 163, 164, 173
Manufacturer dependent 101
Manufacturers
20, 44, 101, 154, 163, 165
Manufacturer's documentation 46
Measurements 154

Mechanical components 23, 27
Mechanical tuners 163
Megahertz (MHz) 11
Memory 176
Menu 174, 176, 179
Metal cabinet parts 27
Metal parts 21, 22, 27
Meter 22
Meter calibration 26
Meter terminals 48
Microcomputer circuits 180
Microfarads 66
Microprocessor
173, 174, 176, 177, 178, 179, 182
Microprocessor circuits 179
Mid-screen 136
Millimeter 45
Misconvergence 106, 198
Mixer
163, 164, 165, 166, 167, 168
Mixer circuit 168
Mixer frequency 113
Mixer stage 163, 165
Mode switch 56
Model 21, 25, 36, 46, 101, 125,
173, 175, 178, 182
Modified sawtooth scan signal 140
Modular tuner 163, 168
Modulated stereo signal (L-R) 114,
115
Modulator 97
Modulator circuit 97
Monaural 9
Monaural audio (L+R) signal 114
Monaural signal (L+R) 115
Monaural sound 116, 117
Monitors 23, 26
Mono 9, 114
MOS (metal oxide semiconductor)
devices 47
Most negative voltage 39
Motor boating 119
Mounting bolt 153
Mounting hardware 20, 25, 27
Movie theater sound 115
Muffled sound 120
Multimeter 43
Multiple contact switch (capacitor)
166
Multiple symptoms 128
Multitester 42
Mute function 115, 175

N

Negative 105
Negative base line 78
Negative cycle alternation 62
Negative picture 198
No color 98, 199
No luminance 199
No picture 104, 199, 203
No SAP 117
No scanning raster 104
No sound 116, 118, 199, 203, 204, 205
No stereo 116, 204
No video 14
Noise 87, 89, 115, 117, 126, 127, 128, 145, 153, 167, 191
Noise protection 45
Noise reduction 73
Noise reduction circuits 115, 117
Non-parallel path 41
Non-stereo televisions 114
Nonlinear picture 199
Nonlinear signal 141
NPN (Negative-Positive-Negative) 37, 39
NPN transistor 40, 72, 75
NTSC (National Television System Committee) 34, 100
NTSC color bar generator 100
NTSC color system 100
NTSC signal generator 116
Numbers 178

O

Off state 73
Off time 74
Ohmmeter 21, 26, 27, 38, 41, 42, 117, 151
Ohms 117
Ohm's law 41
On state 73
On time 74
On-screen display (OSD) 172, 177, 178, 191
On-screen display (OSD) circuits 178
On-screen displays 178, 179
On-screen menu 173
On/off relay circuit 182
Open 40

Open bypass capacitor 40
Open choke 68
Open circuit 174
Open components 56, 157
Open connection 42, 44, 117, 118, 120
Open diodes 145
Open heater element 105
Open primary 77
Open transistor 36, 42, 118
Open-circuited resistor 68
Operating circuit 67
Opposing voltage charges 65
Options 173
Original picture 8
Oscillating output frequency 131
Oscillator 15, 74, 97, 98, 100, 131, 140, 163, 164, 167, 177
Oscillator circuit 100
Oscillator frequency 113, 164
Oscillator pins 178
Oscillator signal 165
Oscillator stage 164
Oscillatory energy 78
Oscilloscope 32, 33, 36, 37, 66, 85, 89, 91, 92, 93, 98, 99, 100, 116, 118, 120, 127, 128, 129, 132, 138, 144, 145, 152, 174, 178, 180, 182
Oscilloscope screen 71
OSD circuits 178, 179, 180
OSD data 179
OSDs 174, 178
Output 10, 36, 37, 39, 58, 67, 69, 98, 102, 105, 110, 116, 117, 130, 131, 132, 140, 143, 151, 153, 155, 158, 162, 165, 168, 177, 179
Output capacitor 68
Output circuit 133, 143
Output connections 119
Output current 72
Output DC voltage 64
Output filter capacitor 75, 76
Output lead 39
Output load 68
Output ripple frequency 76
Output signal 15, 16, 39, 99, 138, 163
Output terminal 63

Output transistors 24
Output values 131, 140
Output voltage 58, 62, 68, 69, 71, 72, 73, 91, 129, 132, 138, 144
Overloaded circuit 138
Overshoot 94
Oxidation 48
Oxidation buildup 105
Oxide surface 103
Ozone 36

P

Paper tube 153
Parallel 27, 41
Parallel component 36
Parallel path 41
Parasitics 45
Part substitution 41
Parts 21, 37
Parts list 23, 27
Passive components 40, 173
Peak inverse voltage (PIV) 61
Peaked sawtooth 143
Peaking coil 94
Peaks 68
Performance 45
Phase 177
Phase difference 129
Phase lock 100
Phase locked loop (PLL) 166
Phase relationships 177
Phase shift 99
Phosphor screen 102
PHOTOFACT 26, 101
PHOTOFACT Safety Precautions 26
Photograph 6
Picture 6, 8, 9, 10, 35, 37, 41, 86, 87, 88, 89, 94, 95, 98, 99, 101, 116, 119, 124, 129, 131, 132, 136, 137, 140, 141, 145, 152, 155, 157, 167, 180, 192, 199, 204
Picture adjustment controls 92
Picture brightness 91
Picture control circuits 95, 180
Picture control settings 93
Picture controls 92, 93, 95, 174
Picture dimming 154
Picture distortions 141

INDEX

Picture height 16, 144
Picture information 13
Picture instability 90
Picture interference 164
Picture linearity 141
Picture loss 14
Picture problem 116, 126, 137, 144, 204
Picture quality 7
Picture quality controls 172
Picture rolling 127, 198
Picture scan rate 15
Picture screen 140
Picture settings 93
Picture signals 7
Picture symptoms 194
Picture tearing 198
Picture tube 7, 33, 104, 105, 132
Picture tube socket 105
Picture width problem 137
Picture-in-picture (PIP) 101, 200
Pilot demodulation circuit (L-R) 116
Pilot signal 114, 115
Pin 93, 151, 173, 175, 180
Pincushion circuit 141
Pincushion correction circuit 136
Pincushioning 136
Pinched wires 27
Pincher cutter 47
Pincushioning 196
Pins 105, 125, 167
PIP 84, 92, 101 (*See Picture-in-picture*)
PIP circuit 92, 101, 102
PIP circuit input 102
PIP processing circuits 102
PIP processing IC 102
PIP window 101
Placement 14
Plastic knobs 22
Plug 105
Plug prongs 21, 27
Plug-in module 101
PNP (Positive-Negative-Positive) 37, 75
PNP transistor 40
Polarity 39, 40, 43, 46, 62, 73, 75, 88, 129, 136
Poor reception 166
Popping 119, 202
Port 82
Positive 39, 40

Positive charge 103
Positive cycle alternation 62
Positive electrodes 103
Positive half-cycles 63
Positive ions 65
Positive terminal 70
Potentiometer 92, 93, 145, 165
Powdered iron cores 76
power 77, 133, 182
Power circuits 22
Power components 39
Power cords 22, 67
Power dissipation 73
Power line 62
Power line cord 66
Power line plug 66
Power loss 176
Power outages 176
Power output 73
Power pin 182
Power rating 71
Power receptacle 22, 26
Power resistors 24
Power source 26, 64, 82
Power spikes 39
Power supply 20, 36, 56, 66, 119, 150, 165, 166, 167, 168, 173, 182, 192
Power supply circuit 81
Power supply problems 67
Power supply stages 56, 150
Power supply voltage 68, 87
Power surges 39, 40
Power switch 21, 66
Power transformers 68, 78
Power transistors 39, 143
Preamplifier 180, 182
Presence decoder 166
Primary 58, 59
Primary colors 97
Primary voltage 59, 60
Primary windings 152
Probe wires 47
Probes 48
Procedures 22, 26
Process 10
Processing circuits 92
Program 173, 174, 180
Program sound 86
Programming 173
PROM (programmable read-only memory) 173

Protection circuits 23, 27, 154
Pulsating DC 57, 62, 63
Pulsating DC output 64
Pulse 78, 140
Pulse rectification 78
Pulse voltage 78
Pulse-width modulator (PWM) 74
Pure color 99
Purity adjustment 106
Purity magnets 107
Push-pull amplifier 131
Push-pull transistors 145
Putt-putt 119, 203

Q

Quartz crystal circuit 177
Quartz tuning 166

R

R-B static convergence magnet 107
R-G-B magnets 107
Radiation 163
Radio 6
Radio frequency (RF) 10 (*See RF*)
Radio frequency signals 8, 45
RAM (random-access memory) 173, 176, 177
RAM circuit 176, 177
Range 155
Raster 8, 87, 90, 91, 116, 156, 157, 167, 168, 192, 203, 204
Raster pattern 167
Raster scanning pattern 179
Rated value 23, 26, 41
Reactance stage 129
Reactance tuner 166
Rear speakers 115
Receiver
  7, 8, 9, 20, 21, 22, 23, 26, 27, 33, 39, 106, 114, 182
Receiving signals 9, 10
Recovery time 76
Rectification 56, 57, 60, 64, 73, 75
Rectified voltage drop 69
Rectifier 72, 78, 79
Rectifier circuit 63
Rectifier tester 66
Recycling 25

Red 8, 97, 105
Red beams 107
Red guns 106
Red signals 179
Reduced brightness 202
Reduced focus 105
Reference signal 129, 164
Reference voltage 69, 72
Reference voltage generator 69
Reference-voltage source 74
Regulated output 76
Regulated power supplies 68
Regulation 22, 73, 74
Regulator 68, 69, 70, 73
Regulator circuit 68, 69
Regulator control loop action 72
Regulator input 70
Regulator output 70
Relay 75
Remote 175, 179
Remote control
  56, 174, 175, 176, 179,
  180, 182, 193
Remote-in pin 180
Remote-controlled volume 116
Repair documentation 9
Repair technician 26
Replacement 168
Replacement parts 20, 21, 23, 26
Replacement transistor 38
Reproduced picture 16
Reset 173, 174, 175, 193
Reset circuit 173, 174
Reset command 174
Reset pin 173, 174, 175
Reset signal 174
Resetting 177
Resistance 26, 27, 33, 37, 40,
  41, 44, 45, 61, 74, 95, 117,
  118, 119, 152, 175
Resistance arrays 45
Resistance change 118
Resistor 20, 21, 22, 25, 26, 27,
  36, 39, 40, 41, 42, 64, 66, 67,
  68, 119, 132, 133, 143, 175
Resistor array 41
Resistor/capacitor combination 177
Resolution 7
Resonant 134
Resonant circuit 94
Resonant frequency 135

Retrace 96, 134, 135, 136,
  139, 140, 143
Retrace cycle 140
Retrace lines 8, 96, 152
Retrace sequence 135
Retrace time 78
Retune 162
Return path 21, 27
Reverberation 115
Reverse biased 39
Reverse breakdown 71
Reverse voltage 62
Reverse-biased 60
Reverse-biased diode (VR) 61
Reverse-biased mode 70
RF 10
RF (radio frequency) signal 10
RF amplifier 163, 164, 167
RF amplifier circuit 164
RF amplifier stage 164
RF frequency 162
RF signal 12, 162, 164, 165
RF switching 172
RF voltages 85
RGB signals 179
Right channel 115
Right signals 114
Right-to-left instructions 7
Ringing 204
Ringing 152
Ringing sound 137
Ripple 64, 73, 76
Ripple voltage 57, 64
Ripple voltage filtering 75
Rolling 7, 90
ROM (read-only memory)
  173, 177, 179
Routing 154
Rubber pads 22
Rubber wedges 106, 107
Running sync 116, 203

S

Safety 20, 26, 59
Safety check 20, 27, 36
Safety glasses 25
Safety guidelines 20, 150
Safety information 20
Safety precautions 26
Safety problems 26
Safety requirements 21
Safety resistor 77

Safety symbols 26
Safety tips 26
Sampling circuit 69
Sampling element 74
SAP (second audio program) 114
SAP processing circuits 117
SAP signal 115, 116
Satellite dish 10
Satellites 10
Saturate 76
Sawtooth generator 140
Sawtooth output signal 140
Sawtooth signal 140, 141, 143
Sawtooth waveform 140, 152
Scan 135, 136, 139, 143, 167
Scan line 7, 134, 135
Scan rectification 78
Scan sequence 134
Scan-derived circuits 79, 82
Scan-derived power supplies
  78, 79, 137
Scan-derived signal pulse 78
Scan-derived signal waveform 78
Scan/retrace sequence 135, 136
Scanning pattern 7, 8, 167, 168
Scanning raster 8, 86, 87, 89, 94,
  104, 124
Schematic 23, 27, 35, 36, 37, 39,
  41, 42, 46, 56, 65, 84, 113,
  119, 125, 131, 135, 140, 151,
  153, 154, 166, 173, 175, 178,
  180, 182
Schematic symbols 60
Schottky diode 76
SCR (silicon control rectifier) 131,
  136
Scratches 25
Screen 33, 96, 97, 101, 103,
  106, 124, 128, 131, 135,
  139, 140, 144, 179, 180, 199
Screen spots 194
Seals 25, 107
Second anode 102, 157
Second anode button 102
Second anode voltage 155
Second audio program (SAP) 203
  (See SAP)
Second stage 86
Secondary 58, 59, 63
Secondary output voltage 63
Secondary peak voltage, VSpk 64

INDEX

Secondary power supplies 79
Secondary voltage 59, 60, 63, 64
Secondary windings 59, 79, 152
Self-oscillating 72
Semiconductor diodes 70
Semiconductors 46
Sensing circuit 76
Series 41
Series resistive fuse 66
Series-pass (linear) 72
Series-pass control element 74
Series-pass feedback voltage regulator 72, 76
Series-pass power supply 72
Service literature 92
Service technicians 20
Set phase relationship 177
Shadow mask 102, 105
Shadows 106
Sharpness 180
Sharpness control 94, 95
Shatterproof goggles 22, 26
Shields 20, 27, 70
Shock 24, 48, 59
Shock hazard 22, 27
Shock waves 47
Short 76, 91, 119, 132, 152, 157, 158
Short picture 144
Shorted capacitor 132
Shorted component 173
Shorted connections 120
Shorted diodes 82
Shorted transistor 39
Shunt regulator tube 23
Shutdown circuit 155
SIF circuit 92
Signal 10
Signal 6, 8, 10, 13, 36, 37, 39, 40, 68, 77, 85, 87, 89, 90, 91, 92, 95, 97, 100, 101, 102, 106, 113, 114, 115, 116, 118, 124, 125, 128, 131, 132, 139, 140, 141, 144, 145, 162, 163, 164, 165, 175, 176, 177, 180, 182
Signal activity 178
Signal detection 88
Signal frequency 113
Signal gain 95
Signal generator 15, 118, 132, 152

Signal injection 37, 89, 117, 130, 132
Signal input waveform 102
Signal path 45
Signal pulse 136
Signal source 34, 100
Signal strength 12
Signal tracing 37, 39, 82, 118, 119, 130, 153
Signal voltage 89, 95
Signal-to-noise-ratio 163
Silicon 37, 39, 61
Silicon controlled rectifier (SCR) 43
Silicon diode 61
Silicon diode replacement 67
Sine-wave voltage 62
Sleep timer 177
Slow motion 101
Small picture 66
Smells 36
SMPS 76, 77
SMPS circuits 77
SMT 44
SMT boards 45
SMTs 46, 47
Snow 8, 167
Solder 46, 47, 153, 154, 168
Solder connections 36, 44, 119, 125, 173, 174
Solder fragments 46
Solder joints 20, 23, 44, 82
Solder splashes 20, 27
Soldered connections 20, 27, 131, 168
Soldering 47
Soldering gun 46, 47
Soldering iron 46, 47, 154
Solid-state circuits 23, 36
Solid-state components 46, 48
Solid-state devices 46, 47, 67
Solid-state receivers 23, 26
Sound 6, 9, 10, 36, 86, 87, 90, 113, 116, 118, 119, 132, 167, 168, 192, 204
Sound alignment 116, 119
Sound basics 9
Sound carrier 14
Sound detector 14
Sound distortions 119
Sound frequency 113
Sound IF 13

Sound IF amplifier 13, 14, 89, 113
Sound IF detector 113
Sound IF signal 13, 89
Sound IF stage 113
Sound levels 175
Sound problem 116
Sound processing system 14
Sound quality 204
Sound signals 7
Sound system 116
Sources 82
Speaker schematic 117
Speakers 14, 115, 116, 117, 118, 120
Special adjustment tool 120
Specifications 117
Spot 105
Square 8
Squares of light 6
Squealing 119, 153, 205
Stability 73
Stages 9, 39, 50, 90, 91, 157, 162
Standby power supply 56, 76, 77, 174, 182
Standby state 73
Startup power supplies 79
Static 36
Static charge 23, 166
Static charge buildup 24
Static convergence 105, 106
Static discharge rug 23, 166
Static electricity 47
Static high voltage 22, 26
Stations 90
Step generator 166
Step-down regulator 75
Step-down switching power supply 75
Step-down transformer 60
Step-up regulator 75
Stepped control circuit 175
Stepped controlled 175
Stepped voltage 173
Stepped voltage control 93
Stereo 9, 114, 116, 117, 175
Stereo demodulator 114
Stereo demodulator path 116
Stereo processing IC 117
Stereo signal 114, 117
Stereo-decoder mode 172

Stray resistance 41
Sub-contrast 95
Sub-picture level controls 95
Substituting parts 41
Substitution book 37
Supply voltage 70, 102, 124, 129, 131, 140, 145, 167
Supply voltage circuit 167
Support circuits 139
Surface 25
Surface-mount technology (SMT) 44
Surround sound 115
Surround sound processing system 115
Switch 39, 43, 66, 136, 166, 173
Switched-mode 73
Switched-mode power supplies (SMPS) 70, 72, 73, 75, 76
Switched-mode regulator 76
Switched-mode system 74
Switching circuit 33, 92, 133
Switching component 136
Switching device 133
Switching power semiconductor 75
Switching power supply 73
Switching regulator 72, 75
Switching systems 73, 74
Switching time 76
Switching transformer 77
Switching transistor 77
Switching-frequency ripple 76
Symptoms 66, 88, 104, 116, 137, 144, 154, 157, 188
Sync 7, 14, 64, 99
Sync amplifier 14, 126
Sync circuit 116
Sync circuit IC 129
Sync problems 144
Sync pulses 90, 139
Sync sample 178, 180
Sync separator
  13, 14, 89, 124, 125, 126, 128, 129, 139, 145
Sync separator circuit 125, 128
Sync separator circuit IC 129
Sync signal
  10, 125, 128, 129, 139, 191
Sync signal processor 102
Sync signal pulses 85, 89, 129

Sync signals 7, 12, 13, 14, 15, 16, 85, 90, 112, 124, 125, 126, 145
Synchronization 6, 7, 90, 92
Synchronizing signals 7
Synchronous detector 88
System circuits 68
System control 92, 115, 173
System control circuits 92, 116, 173, 174, 177, 180, 182, 183
System control functions 178
System control IC
  93, 114, 172, 173, 179, 182

T

Tearing 7, 14
Tearing pictures 90
Technicians 20
Techniques 32
Television 16, 20
Television basics 6
Television camera 7
Television picture 7
Television picture tube (CRT) 102 (*See CRT*)
Television receiver 7
Television screen 124
Television station
  6, 7, 9, 10, 11, 14, 16, 125, 129, 132, 162, 179
Television studio 97
Temporary storage element inductor 74
Terminal 44, 46, 153
Terminal pin 44
Test 22, 32, 41, 48
Test equipment 22
Test lead 22, 26
Test signal 116, 117
Testing components 42
Thermal overload circuit 66, 68
Third stage 86
Third transistor 89
Three-stage video amplifier 90
Three-terminal IC regulators 68
Ticking 153
Ticking sound 193
Time constant 132
Timing 16, 98, 177
Timing capacitors 134
Timing circuit 131, 140

Tint 98, 180
Tint control 99
Tone 117
Tone control 14
Tone tests 117
Tools 22
Traces 174
Tracking 204
Transformation
  56, 57, 59, 64, 73, 75
Transformer 39, 57, 58, 59, 60, 63, 66, 133, 142, 143, 151, 153, 154
Transformer secondary output 62
Transformer terminals 154
Transformer windings 152
Transient voltage 70
Transistor 33, 37, 38, 39, 40, 41, 42, 43, 46, 56, 67, 75, 77, 85, 89, 91, 99, 100, 118, 119, 130, 133, 137, 139, 143, 145, 158, 164, 173, 174
Transistor amplifier 92
Transistor tester 42
Transmit 11, 162
Transmitted picture 16
Transmitted signal 37
Transmitter 7
Transmitting frequency 10
Treble 175
Triggering device 135
Troubleshooting 32, 39, 86, 182, 188
Troubleshooting methods 37
Tuner 8, 10, 11, 12, 85, 134, 166, 167, 168, 172, 176, 177
Tuner circuits 162, 163, 166, 168
Tuner control circuits 177
Tuner frequency 166
Tuner operations 173
Tuner shield 168
Tuner stage 162, 163
Tuner types 165
Tuners 37, 163, 164, 165
Tuning 162
Tuning controls 164
Tuning varactors 71
Turns ratio 59
Tweezers 46

INDEX

U

UHF channels 10, 11
UHF frequency band 166
Undervoltage lock circuit 76
Uneven vertical deflection 141
Universal replacement components 45, 46
Unmodulated reference signal 164
Unregulated DC power supply 57
Unregulated power supply 59, 60, 68, 77
Unsolder 152
Unstable picture 204
Unused channel 8
Upward tearing 131
Utility line 59
Utility power lines 57

V

Vacuum 103
Values 113, 117, 151
Varactor (variable capacitance diode) 163
Varactor tuner 165
Varactor tuners 163
Variable AC transformer 23, 27
Variable current limited power supply 77
Variable DC supply 93
Variable frequency oscillator (VFO) 177
Variable line transformer 78, 82, 152, 155
Variable oscillator 173, 177
Variable resistance 69
Variable voltage supply 165
VCR 92
Vertical blanking 96, 98
Vertical blanking pulse 98
Vertical circuits 140, 141, 145, 178
Vertical controls 144
Vertical deflection 141, 154
Vertical deflection circuits 102, 124, 125, 126, 127, 139, 144, 145
Vertical deflection IC 145
Vertical deflection yoke 16, 139, 140, 143, 145
Vertical height control 144, 145
Vertical hold 127, 144
Vertical hold control 16, 144, 145
Vertical integrator 126, 127
Vertical lines 107
Vertical oscillators 139, 140, 145
Vertical output circuits 139, 140, 143, 145
Vertical output transistors 139
Vertical pincushion circuit 142
Vertical rate 140
Vertical retrace 139, 140, 143
Vertical rolling 7, 15, 66, 102, 128
Vertical scan 16, 124, 125, 139, 140, 143, 144, 200
Vertical scan section 16
Vertical scan stage 16
Vertical scanning waveform 140
Vertical signals 124, 125, 127, 128, 145
Vertical size 140
Vertical stages 36, 144
Vertical sync 127, 198
Vertical sync circuits 178, 180
Vertical sync signal 14, 16, 102, 126, 201
Vertical synchronizing pulses 96
Vertical tearing 129
Vertical timing circuit 140
Vertical yoke coils 142
Vertical-sync signals 7
Vertically 90
Vertically unstable picture 66
VHF channels 10, 11
Vibrating 153
Video 6, 10, 12, 13, 16, 84, 85, 86, 89, 90, 91, 101, 168
Video amplifier 13, 14, 16, 37, 87, 89, 91, 92, 94, 95
Video amplifier circuit 89, 90, 91, 105
Video basics 7
Video carrier 12, 165
Video circuits 84, 119, 167, 172
Video detector 13, 85, 86, 87, 88, 89, 110, 112, 113, 116, 128
Video detector circuit 128
Video frequency 113
Video gain 91
Video IF amplifier 13, 85
Video IF amplifier IC 94
Video IF circuits 95, 112
Video information 13
Video interference 189
Video output stage circuitry 99
Video path 95, 101
Video processing circuits 92, 95, 125, 167, 179, 180
Video processing IC 92, 96
Video processing stage 92, 157, 166
Video product 36
Video signal 8, 10, 13, 16, 37, 87, 92, 93, 102, 109, 112, 114, 124, 125, 126, 132, 179
Video source 92
Video stage 39, 84
Video timing 14
Video transmission 9
Viewing screen 102
Voice coil 118
Volt 27
Volt-ohmmeter (VOM) 21, 32, 33 (*See VOM*)
Voltage 16, 21, 22, 23, 27, 33, 35, 39, 40, 41, 43, 44, 46, 56, 61, 62, 65, 68, 69, 71, 72, 73, 75, 76, 77, 78, 82, 84, 85, 86, 87, 91, 93, 100, 102, 104, 105, 112, 117, 118, 119, 124, 125, 129, 135, 136, 151, 152, 153, 155, 157, 164, 165, 173, 174, 175, 179, 180, 182
Voltage amplifier 118
Voltage analysis 77
Voltage arcing 23, 24, 48
Voltage changes 71
Voltage discharge 134, 145
Voltage divider 95
Voltage drop 40, 61, 69, 71, 137
Voltage generator 130
Voltage input 174
Voltage leaking 164
Voltage level 12, 68, 175, 179, 180
Voltage measurements 22, 27, 40
Voltage output 69
Voltage path 153, 167
Voltage pulses 137
Voltage range 153, 175
Voltage rating 71
Voltage readings 67
Voltage reduction 136
Voltage regulator circuit 79
Voltage requirements 182

Voltage stabilizers 71
Voltage surges 46
Voltage values 57, 131, 135, 175
Voltage variations 68
Voltmeter 33, 40, 56, 72, 77,
  85, 87, 89, 91, 99, 118
Volume
  87, 117, 119, 167, 175, 176
Volume control
  14, 86, 87, 89, 115,
  116, 117, 118, 173, 202
Volume control pin 175
Volume control problems 205
Volume level 173
VOM 66, 150
VON 75
VR (reverse-bias voltage) 62

W

Wall outlet 153
Warnings 20
Wave 78
Waveform 33, 37, 66, 78, 91,
  96, 99, 100, 118,
  120, 127, 128, 129, 132, 144,
  145, 152
Waveform output 128
Weak picture 90, 202
Weak sound 118, 205
Weak video 86
Weaving 129
Weaving picture 199
Whistling 116, 119, 205
White vertical line 200
Width problems 199
Winding 68, 152, 154
Wire 20, 22, 24, 27, 44,
  58, 76, 118
Wire cutter 47
Wiring problems 119
Work surface 22

X

X-ray radiation
  23, 26, 27, 35, 154
X-rays 23, 26
X10 probe 178

Y

Yoke 106, 107, 135, 136

Z

Zener diode 70, 71
Zener diode symbol 71

PROMPT® Publications

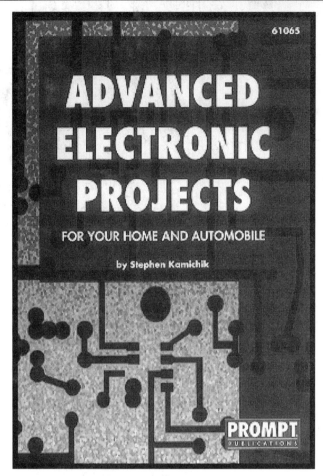

Advanced Electronic Projects for Your Home and Automobile
by Stephen Kamichik

You will gain valuable experience in the field of advanced electronics by learning how to build the interesting and useful projects featured in *Advanced Electronic Projects*. The projects in this book can be accomplished whether you are an experienced electronic hobbyist or an electronic engineer, and are certain to bring years of enjoyment and reliable service.

$18.95
Paper/160 pp./6 x 9"/Illustrated
ISBN#: 0-7906-1065-5
Pub. Date 5/95

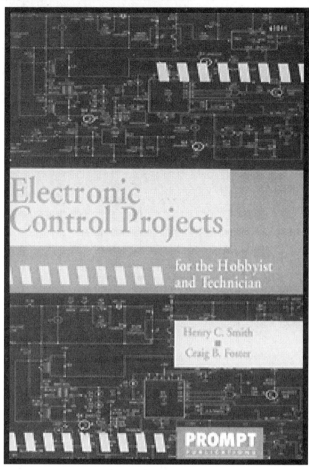

Electronic Control Projects for the Hobbyist and Technician
by Henry C. Smith and Craig B. Foster

Would you like to know how and why an electronic circuit works, and then apply that knowledge to building practical and dependable projects that solve real, everyday problems? Each project in *Electronic Control Projects* involves the reader in the actual synthesis of a circuit. A complete schematic is provided for each circuit, along with a detailed description of how it works, component functions, and troubleshooting guidelines.

$16.95
Paper/168 pp./6 x 9"/Illustrated
ISBN#: 0-7906-1044-2
Pub. Date 11/93

Call us today for the name of your nearest distributor. **800-428-7267**

PROMPT® Publications

Tube Substitution Handbook
Complete Guide to Replacements
for Vacuum Tubes and Picture Tubes
by William Smith and Barry Buchanan

The most accurate, up-to-date guide available, the *Tube Substitution Handbook* is useful to antique radio buffs, old car enthusiasts, ham operators, and collectors of vintage ham radio equipment. In addition, marine operators, microwave repair technicians, and TV and radio technicians will find the *Handbook* to be an invaluable reference tool. Diagrams are included as a handy reference to pin numbers for the tubes listed in the *Handbook*.

$16.95
Paper/ 149 pp./6 x 9"/Illustrated
ISBN#: 0-7906-1036-1
Pub. Date 12/92

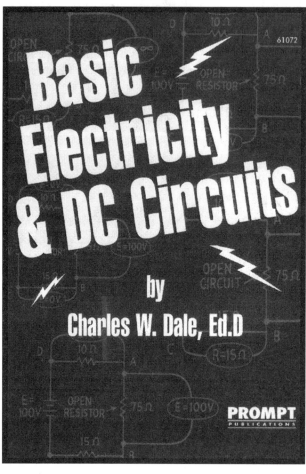

Basic Electricity & DC Circuits
by Charles W. Dale, Ed.D

Electricity is constantly at work around your home and community, lighting rooms, running manufacturing facilities, cooling stores and offices, playing radios and stereos, and computing bank accounts. Now you can learn the basic concepts and fundamentals behind electricity and how it is used and controlled. *Basic Electricity and DC Circuits* shows you how to predict and control the behavior of complex DC circuits. The text is arranged to let you progress at your own pace, and the concepts and terms are introduced as you need them, with many detailed examples and illustrations.

$34.95
Paper/928 pp./6 x 9"/Illustrated
ISBN#: 0-7906-1072-8
Pub. Date 8/95

800-428-7267 Call us today for the name of your nearest distributor.

PROMPT® Publications

Industrial Electronics for Technicians

by J. A. Sam Wilson and Joseph Risse

Industrial Electronics for Technicians provides an effective overview of the topics covered in the Industrial Electronics CET test, and is also a valuable reference on industrial electronics in general. This workbench companion book covers the theory and application of industrial hardware from the technician's perspective, giving you the explanations you need to understand all of the areas required to qualify for CET accreditation.

$16.95
Paper/352 pp./6 x 9"/Illustrated
ISBN#: 0-7906-1058-2
Pub. Date 8/94

Mark Waller's Harmonics

A Field Handbook for the Professional and the Novice

Many operational problems can be solved through an understanding of power system harmonics. As life/safety issues become more and more important in the world of electrical power systems, the need for harmonic analysis becomes ever greater. This book is the essential guide to understanding all of the issues and areas of concern surrounding harmonics and the recognized methods for dealing with them.

$24.95
Paper/132 pp./7-3/8 x 9-1/4"
ISBN#: 0-7906-1048-5
Pub. Date 3/94

Call us today for the name of your nearest distributor. **800-428-7267**

𝄞 𝄞 PROMPT® Publications 𝄞 𝄞

Speakers for Your Home and Automobile
How to Build a Quality Audio System
by Gordon McComb, Alvis J. Evans, and Eric J. Evans

The cleanest CD sound, the quietest turntable, and the clearest FM signal are useless without a fine speaker system. With easy-to-understand instructions and illustrated examples, this book shows how to construct quality home speaker systems and how to install automotive speakers.

$14.95
Paper/164 pp./6 x 9"/Illustrated
ISBN#: 0-7906-1025-6
Pub. Date 10/92

Sound Systems for Your Automobile
The How-To Guide for Audio System Selection and Installation
by Alvis J. Evans and Eric J. Evans

Whether you're starting from scratch or upgrading, this book will show you how to plan your car stereo system, choose components and speakers, and install and interconnect them to achieve the best sound quality possible. Easy-to-follow steps, parts lists, wiring diagrams, and fully illustrated examples make planning and installing a new system easy.

$16.95
Paper/124 pp./6 x 9"/Illustrated
ISBN#: 0-7906-1046-9
Pub. Date 1/94

800-428-7267
Call us today for the name of your nearest distributor.

PROMPT® Publications

Advanced Speaker Designs for the Hobbyist and Technician
by Ray Alden

This book shows the electronics hobbyist and the experienced technician how to create high-quality speaker systems for the home, office, or auditorium. Every part of the system is covered in detail, from the driver and crossover network to the enclosure itself. You can build speaker systems from the parts lists and instructions provided, or you can actually learn to calculate design parameters, system responses, and component values with scientific calculators or PC software.

$16.95
Paper/136 pp./6 x 9"/Illustrated
ISBN#: 0-7906-1070-1
Pub. Date 7/94

Security Systems for Your Home and Automobile
by Gordon McComb

Because of the escalating threat of theft and violence in today's world, planning, selecting, and installing security systems to protect your home and automobile is vital. You can save money by installing a system yourself. In simple, easy-to-understand language, *Security Systems for Your Home and Automobile* tells you everything you need to know to select and install a security system with a minimum of tools.

$16.95
Paper/130 pp./6 x 9"/Illustrated
ISBN#: 0-7906-1054-X
Pub. Date 7/94

Call us today for the name of your nearest distributor.   **800-428-7267**

PROMPT® Publications

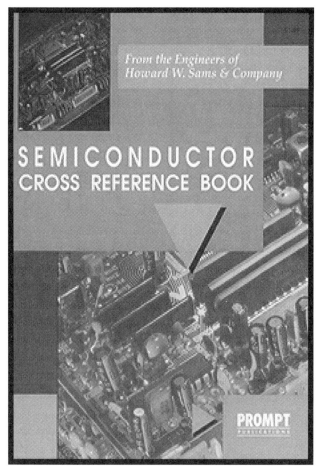

Semiconductor Cross Reference Book
Revised Edition
by Howard W. Sams & Company

From the makers of PHOTOFACT® service documentation, the *Semiconductor Cross Reference Book* is the most comprehensive guide to semiconductor replacement data. The volume contains over 475,000 part numbers, type numbers, and other identifying numbers. All major types of semiconductors are covered: bipolar transistors, FETs, diodes, rectifiers, ICs, SCRs, LEDs, modules, and thermal devices.

$24.95
Paper/668 pp./8-1/2 x 11"
ISBN#: 0-7906-1050-7
Pub. Date 6/94

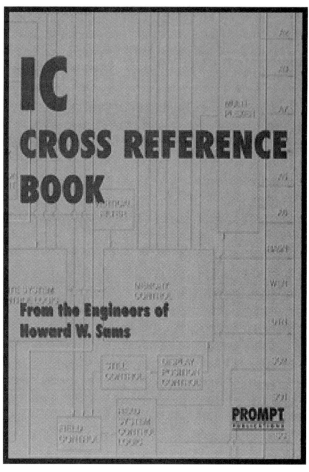

IC Cross Reference Book
by Howard W. Sams & Company

The engineering staff of Howard W. Sams & Company has assembled the *IC Cross Reference Book* to help you find replacements or substitutions for more than 35,000 ICs or modules. It has been compiled from manufacturers' data and from the analysis of consumer electronics devices for PHOTOFACT® service data, which has been relied upon since 1946 by service technicians worldwide. This unique book includes a complete guide to IC and module replacements and substitutions, an easy-to-use cross reference guide, listings of more than 35,000 part and type numbers, and more.

$19.95
Paper/168 pp./8-1/2 x 11"
ISBN#: 0-7906-1049-3
Pub. Date 5/94

800-428-7267 Call us today for the name of your nearest distributor.

PROMPT® Publications

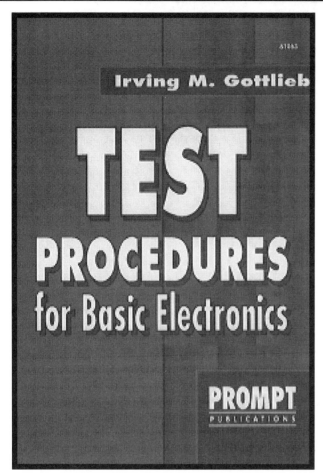

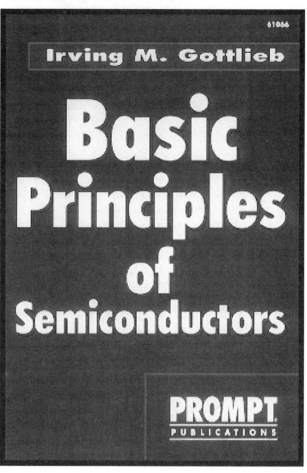

Test Procedures for Basic Electronics
by Irving M. Gottlieb

Test Procedures for Basic Electronics covers many useful electronic tests and measurement techniques, with emphasis on the use of commonly available instruments. Students, hobbyists, and professionals are provided with the whats and whys of obtaining useful results, whether they are repairing a modern CD player or restoring an antique radio.

$16.95
Paper/356 pp./7-3/8 x 9-1/4"
ISBN#: 0-7906-1063-9
Pub. Date 12/94

Basic Principles of Semiconductors
by Irving M. Gottlieb

Few books offer the kind of concise and straightforward discussion of electrical concepts that is found in *Basic Principles of Semiconductors*. From an exploration of atomic physics right through a detailed summary of semiconductor structure and theory, the reader goes on a step-by-step journey through the world of semiconductors. Some of the subjects covered include electrical conduction, transistor structure, power MOSFETs, and Gunn diodes.

$14.95
Paper/161 pp./6 x 9"/Illustrated
ISBN#: 0-7906-1066-3
Pub. Date 4/95

Call us today for the name of your nearest distributor.   **800-428-7267**

PROMPT® Publications

Introduction to Microprocessor Theory & Operation
A Self-Study Guide With Experiments
by J.A. Sam Wilson and Joseph Risse

Introduction to Microprocessor Theory & Operation takes you into the heart of computerized equipment and reveals how microprocessors work. By covering digital circuits in addition to microprocessors and providing self-tests and experiments, this book makes it easy for you to learn microprocessor systems.

$16.95
Paper/212 pp./6 x 9"
ISBN#: 0-7906-1064-7
Pub. Date 2/95

Schematic Diagrams
The Basics of Interpretation and Use
by J. Richard Johnson

Step-by-step, *Schematic Diagrams* shows you how to recognize schematic symbols and their uses and functions in diagrams. You will also learn how to interpret diagrams so you can design, maintain, and repair electronics equipment. Subjects covered include component symbols and diagram formation, functional sequence and block diagrams, power supplies, audio system diagrams, computer diagrams, and more.

$16.95
Paper/208 pp./6 x 9"/Illustrated
ISBN#: 0-7906-1059-0
Pub. Date 9/94

800-428-7267 Call us today for the name of your nearest distributor.

PROMPT® Publications

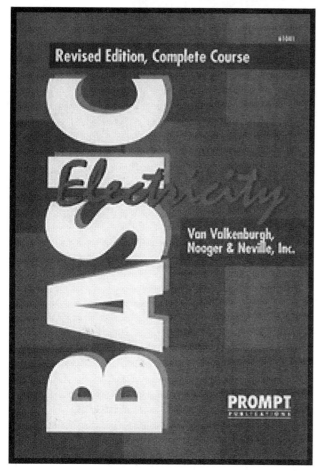

Basic Electricity
Revised Edition, Complete Course
by Van Valkenburgh, Nooger & Neville, Inc.

From a simplified explanation of the electron to AC/DC machinery, alternators, and other advanced topics, *Basic Electricity* is the complete course for mastering the fundamentals of electricity. The book provides a clear understanding of how electricity is produced, measured, controlled, and used. A minimum of mathematics is used in the direct explanation of primary cells, magnetism, Ohm's law, capacitance, transformers, DC generators, AC motors, and other essential topics.

$19.95
Paper/736 pp./6 x 9"/Illustrated
ISBN#: 0-7906-1041-8
Pub. Date 2/93

Basic Solid-State Electronics
Revised Edition, Complete Course
by Van Valkenburgh, Nooger & Neville, Inc.

Modern electronics technology manages all aspects of information—generation, transmission, reception, storage, retrieval, manipulation, display, and control. A continuation of the instruction provided in *Basic Electricity*, *Basic Solid-State Electronics* provides the reader with a progressive understanding of the elements that form various electronic systems. Electronic fundamentals covered in the illustrated, easy-to-understand text include semiconductors, power supplies, audio and video amplifiers, transmitters, receivers, and more.

$24.95
Paper/944 pp./6 x 9"/Illustrated
ISBN#: 0-7906-1042-6
Pub. Date 2/93

Call us today for the name of your nearest distributor. **800-428-7267**

PROMPT® Publications

Power Supplies
Projects for the Hobbyist and Technician
by David Lines

Power supplies, the sources of energy for all electronic equipment, are basic considerations in all electronic design and construction. This book guides you from the fundamentals of power supply components and their functions to the design and construction of a power supply system.

$10.95
Paper/96 pp./6 x 9"/Illustrated
ISBN#: 0-7906-1024-8
Pub. Date 12/92

Simplifying Power Supply Technology
by Rajesh J. Shah

Simplifying Power Supply Technology is an entry point into the field of power supplies. It simplifies the concepts of power supply technology and gives the reader the background and knowledge to confidently enter the power supply field.

$16.95
Paper/160 pp./6 x 9"/Illustrated
ISBN#: 0-7906-1062-0
Pub. Date 3/95

800-428-7267 Call us today for the name of your nearest distributor.

PROMPT® Publications

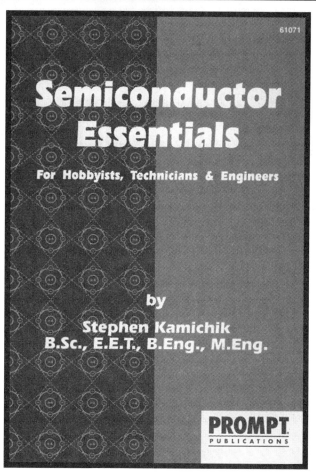

Surface-Mount Technology for PC Boards
by James K. Hollomon, Jr.

Manufacturers, managers, engineers, and others who work with printed-circuit boards will find a wealth of information about surface-mount technology (SMT) and fine-pitch technology (FPT) in this book. Practical data and clear illustrations plainly present the details of design-for-manufacturability, environmental compliance, design-for-test, and quality/reliability for today's miniaturized electronics packaging.

$26.95
Paper/309 pp./7 x 10"/Illustrated
ISBN#: 0-7906-1060-4
Pub. Date 7/95

Semiconductor Essentials
by Stephen Kamichik

Gain hands-on knowledge of semiconductor diodes and transistors with help from the information in this book. *Semiconductor Essentials* is a first course in electronics at the technical and engineering levels. Each chapter is a lesson in electronics, with problems presented at the end of the chapter to test your understanding of the material presented. This generously illustrated manual is a useful instructional tool for the student and hobbyist, as well as a practical review for professional technicians and engineers.

$16.95
Paper/112 pp./6 x 9"/Illustrated
ISBN#: 0-7906-1071-X
Pub. Date 9/95

Call for the complete list of new releases from PROMPT®. **800-428-7267**

PROMPT PUBLICATIONS

New Release
Available April 1996

The In-Home VCR Mechanical Repair & Cleaning Guide
Curt Reeder

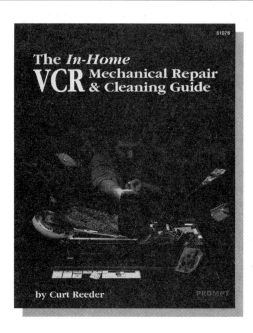

PROMPT® Publications is pleased to announce the release of *The In-Home VCR Mechanical Repair & Cleaning Guide*, a book geared towards the average home VCR user who would like assistance in maintaining their equipment.

Like anything that is used in the home, whether it be a vacuum cleaner, lawn mower, or toaster, a VCR requires minimal service to keep functioning well. A technical or electrical engineering degree is not required to begin maintaining VCRs on a regular basis. With a few small hand tools such as tweezers, cleaning fluid, a power screwdriver and cotton swabs, *The In-Home VCR Mechanical Repair & Cleaning Guide* shows readers the many tricks and secrets of VCR maintenance.

This book is also geared toward the entrepreneur who may consider starting a new VCR service business of their own. The vast information contained in this guide gives a firm foundation on which to create a personal niche in this unique service business.

Curt Reeder started his in-home VCR repair and cleaning service in 1990, servicing VCRs in the homes of his customers. This book is written from the many notes and descriptive remedies he has compiled which illustrate in easy-to-understand terms how an average VCR owner can service their own unit or start a personal, small business. Reeder has repaired literally hundreds of VCRs by using the methods outlines within this text. The most common and frequent VCR malfunctions that Reeder's customers have experienced with their VCRs have brought this book into being.

The In-Home VCR Mechanical Repair & Cleaning Guide
ISBN#: 0-7906-1076-0
Size: 8-3/8 x 10-7/8"
Pages: 222
Price: $19.95
Binding: Paper
Photos: B/W

CALL 1-800-428-7267 FOR MORE INFORMATION

PROMPT® Publications is a division of Howard W. Sams & Company
A Bell Atlantic Company

A broken TV is only as good as the service documentation used to fix it.

In a world of faulty circuits, blown picture tubes, and failed transistors, it's nice to know there's one constant — PHOTOFACT®. For nearly fifty years, *Howard W. Sams & Company* has been producing PHOTOFACT®, and thus setting the industry standard for complete, consistent, and accurate TV repair and service documentation.

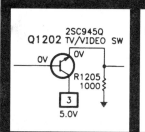

Sams' **PHOTOFACT® design** includes:
- Standard-notation **schematics** drawn by **signal flow**.
- Detailed **electronic parts lists** for each board.
- **Waveforms** and **Voltages**.
- **IC functions** showing internals.

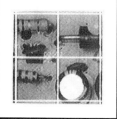

Sams' unique **placement chart** illustrates:
- Element **locations** on each board.
- All **voltage source** locations.
- Every **active component** and its placement.
- Any important **test points** that are found on the boards.

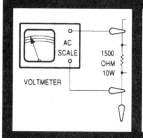

Sams' **gridtrace photos** provide:
- **Quick and easy** location of all parts.
- Exact **photographic representations** of each board.
- Detailed **location guides** listing all elements and their appropriate **coordinates**.

Sams' **miscellaneous adjustments** and **troubleshooting** guides detail:
- Operational **function checks**.
- **Shutdown procedures** to ensure safety.
- Common **troubleshooting tips**.
- **Final adjustment** procedures that require only common, standard **equipment** and **tools**.

PHOTOFACT® can be purchased either individually or by subscribing to the PHOTOFACT-of-the-Month program.

Subscribers receive service documentation on at least 14 different models each month, all for just $49.95.

When you consider that manufacturers' data for 14 models could cost well over $250, it's easy to see what an unbelievable value the PHOTOFACT-of-the-Month program really is.

Call us today to begin your subscription and start saving!

Toll Free Phone:
1-800-428-7267
Toll Free Fax:
1-800-552-3910

Howard W. Sams & Company
Your Information Source for the Electronics Industry
A Bell Atlantic Company

Join the PHOTOFACT®-of-the-Month Club and see how TVs were meant to be fixed.

Subscribe today and you will get the most current PHOTOFACT® folders, covering at least 14 different models and chassis, for only $49.95 per month. If you were to purchase the same PHOTOFACT® individually, they would cost over $165. The savings are even better when you consider how much the original manufacturers' data for all those sets would cost. At today's rate of $15 to $75 per unit, you could easily spend over $250 to get the same coverage that PHOTOFACT®-of-the-Month gives you for only $49.95.

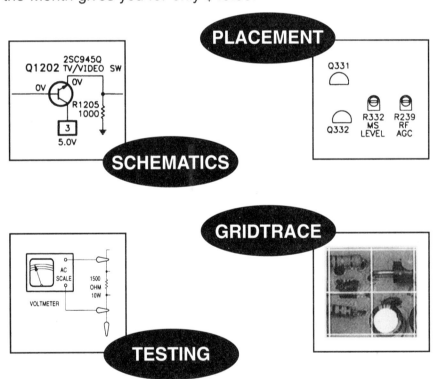

Call today to begin your subscription to the PHOTOFACT®-of-the-Month Club.

1-800-428-7267

Howard W. Sams & Company
A Bell Atlantic Company